201 INDOOR PLANTS IN COLOUR

By the same author

The Complete Book of House Plants

Ideas For Your Garden in Colour

ROB HERWIG

Translated by
Marian Powell

201 INDOOR PLANTS IN COLOUR

LUTTERWORTH PRESS . GUILDFORD AND LONDON

All the photographs in this book were taken in Rob Herwig's studio, with the exception of those showing *Camellia japonica* (taken by Pat Brindley) and *Cytisus x racemosus* (taken by Harry Smith).

Ines Girisch contributed to the preparation of the book, as did the following firms and institutions:

N. V. Handelskwekerij Gebr. Barendsen, Aalsmeer
Bruinsma's Siertuincentrum, Aalsmeer
Cactuskwekerij C. V. Bulthuis en Co., Cothen
Fa. Edelman, Reeuwijk
Instituut voor de Veredeling van Tuinbouwgewassen, Wageningen
W. Lagerwey, Lunteren
Gebr. K. & C. van der Meer, Aalsmeer
Tuinmekka, De Bilt

and others, though all their names could not be given here.

First published in Great Britain, 1976

Second impression, 1980

Dutch edition copyright © 1974 by Zomer & Keuning Boeken B.V., Wageningen

English translation copyright © 1976 by Lutterworth Press

ISBN 0 7188 2263 3

Printed in The Netherlands

CONTENTS

FOREWORD

Many people believe that a plant's sole function is decorative. I don't agree. House plants are living creatures, as responsive to affection, care and interest as we are ourselves. They will reward us by flourishing. You need only observe the contrast between office plants treated with indifference, and the same plants cherished on the window-sill by a true plant lover. Thus, compared with such a hobby as collecting match boxes, plants give something in return.

If you feel as I do, this book has been written for you. I am not interested in the annual amount you spend on house plants: to spend a pound or two and get a lot of pleasure from your plants is better than spending lots of money for show. That is why this book contains a great many hints on bringing house plants through the winter, on sowing and taking cuttings; these are the things which will give meaning to your hobby. To be fully comprehensive, so-called 'disposable plants' are mentioned, but only grudgingly. Fortunately the plants belonging to this category are in the minority.

To begin with, there are ten pages of introductory notes with black and white photographs covering the chief aspects of plant care. These pages will frequently be referred to in the alphabetical section and it would therefore be best for you to read them first.

CHOOSING THE POT

The choice of pot depends on the plant for which it is intended. Plants which like the compost to be kept moist are best grown in a non-porous pot: a plastic one, for instance. If the plant prefers dry compost, a clay pot is better. It is now possible to buy plastic tubs suitable for larger plants (for example, bay trees) which are to be placed outdoors in summer.

As for ornamental pots, I like ones with matching saucers (several of which appear in this book). They allow drainage and thus you can avoid the problem of excess water in the bottom of the pot, which can prove fatal to plants in undrained ornamental pots. Plants in plastic or glazed pots need far less water than those grown in clay pots.

WHICH IS THE BEST COMPOST?

A plant grows best in soil similar to that of its native habitat. As this is not always easy to obtain, standard potting composts have been developed which are obtainable practically everywhere. These are the John Innes potting and seed composts, which contain a balanced mixture of loam, peat and sharp sand, chalk and a compound fertiliser made up of the nutrients supplying nitrogen, phosphorus and potassium. There are three potting composts containing successively greater amounts of fertiliser and chalk, and one seed compost.

There are also the soilless composts, which do not contain loam, but only peat and sand, with or without chalk and fertiliser according to brand. These composts need slightly different watering from the John Innes composts, and are best used for the smaller plants.

7

REPOTTING

Rapidly growing plants, in particular, should be repotted frequently, since the nutrient content of the relatively small volume of compost is quickly exhausted: an abutilon, for instance, will only grow to its full potential if it is potted on into a larger pot three times in a season!

As a rule one is advised to repot plants just before they start into growth again; early in spring, for instance. This is sound advice, but if you are always careful to remove the complete soil ball from the pot, you *can* repot whenever you like, even in the middle of summer. The roots remain intact and the plant is unaffected by the operation. Cutting off old roots or removing old compost, as is still occasionally suggested, is practically superfluous. However, it is important that long roots, which have grown through the drainage hole, are left intact. This can only be ensured by breaking the pot.

The photograph above left shows the soil ball of a potbound aspidistra, which urgently requires repotting. If the soil ball is difficult to remove, invert the pot and knock the rim on the edge of a table. Take the new pot, which should be considerably larger, and place a crock—a broken piece of clay pot—on the bottom, rounded side up, over the drainage hole. Put some fresh compost in the pot, then put the old soil ball carefully in the centre and fill the space around it with new compost. Firm the compost down. Make sure that the soil ball is placed well down in the pot, do not add too much compost, but leave space for watering at the top.

If a plant has grown several crowns, it can be divided into two or three parts. Whenever possible use your hands for this operation (tear the sections apart); only use a knife if absolutely necessary. Pot each section separately.

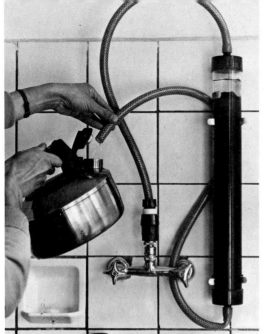

WATERING

If your home is in an area where the tapwater is very hard, it is really unsuitable for your house plants. Azaleas, anthuriums, and heathers, in particular, are very sensitive to hard water. If you live in an area where there is little industrial activity, you can use rainwater collected in a water butt. Let the water reach room temperature before using it on your plants. If, however, you regularly notice a film of oil on the water in the butt, you will have to filter the rainwater before using it. A filter, based on ion exchange, which needs to be washed with salt, is of no use. Complete or partial demineralisation is essential to produce water suitable for your plants. Filters need cost no more than a few pounds. The photograph above left shows a useful filter.

HOW MUCH WATER?

Many people appear to have difficulty in deciding how much water a plant needs. The amount varies considerably.

A cactus or succulent may be left dry for a time; conversely, a begonia likes its compost to be kept constantly moist. Most plants, however, prefer the soil ball to be allowed to dry out almost completely between watering.

The easiest way to determine whether the compost is dry or wet is to test it with your forefinger. A moisture meter, shown in the photograph above right, can be pushed deeper into the pot and is more accurate. The results noted on testing the soil, combined with other factors covered in this book, should make it simple to decide whether the plant needs water.

SOAKING TO REVITALISE

We try to supply our plants' need for moisture by giving them the right amount of water. Nevertheless some plants respond to a brief soaking from time to time, though this should only be of short duration. By submerging the pot in water, air is removed from the soil and part of the accumulated waste matter is washed away. The air is automatically replaced while the plant is draining. Woody plants in particular, hibiscus, abutilon, etc., like this kind of treatment.

The easiest way to immerse the pot is in a sink filled to the pot's rim with lukewarm, and preferably lime-free, water. When no more air bubbles rise to the top of the compost the plant can be removed.

AIR MOISTURE

It is easy to give a plant enough water, but much more difficult to keep the air sufficiently humid. In centrally heated houses especially, the air is too dry for nearly all plants. Humidifiers attached to the radiator are inadequate. An extended window-sill helps to keep the plants away from the very dry air. A plant container filled with moist peat, even if placed above a radiator, does much to improve conditions. It is even better if such a container is surrounded by glass, creating a so-called 'plant window'.

The best method of increasing air moisture is to use electric evaporators or humidifiers, but these are never completely silent. A handspray is cheap and simple, but to be effective, must be used very frequently.

FEEDING YOUR PLANTS

Every growing plant needs nourishment and, as soon as the compost is exhausted (which may take only a few weeks), new food has to be provided; otherwise growth will be retarded and the plant will become susceptible to disease.

Since good potting compost is rich in nutrients, you could repot at frequent intervals instead of feeding. If you find this too much trouble, there are plenty of preparations available, some based on artificial fertiliser, others entirely organic, which can be given to the plant by means of a watering can. Greedy plants, such as gloxinia (sinningia), can be given a nutrient solution almost daily, while a fortnightly feed is sufficient for moderately growing species.

AVOIDING STRONG SUNLIGHT

There are only a few plants, cacti and succulents, for instance, which can tolerate strong sunlight throughout the day. Recommendations with regard to sunlight are given in the following descriptions of individual plants.

For many plants an east-facing window is particularly favourable. A west-facing one is suitable as well, but here it will be colder in the morning. In neither of these positions does the sun shine too long or too strongly, and most plants will tolerate these aspects. In a south-facing window, on the other hand, some measure of screening between 10 a.m. and 4 p.m. is desirable. The best way to regulate the amount of sunshine is by means of Venetian blinds, but even net curtains have a subduing effect, as does an outside sun awning.

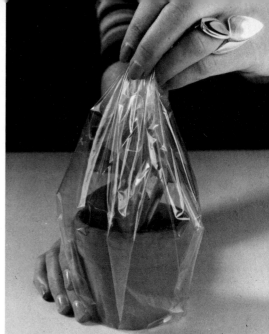

TAKING CUTTINGS

An incredible number of plants can be increased from cuttings, especially if you possess a small indoor propagator in which the soil temperature can be raised to 25–30°C (77–86°F) by means of an electric heating flex or by some other method.

New plants can be grown from shoot tips, sections of roots (root cuttings) and even from stem sections (stem cuttings). The best method for each plant is indicated in the following descriptions, unless, of course, a different type of propagation (e.g., by seed) is preferable.

When taking tip cuttings, the most usual form of cuttings, young shoots are generally best, but in some cases slightly older shoots will strike more easily. Always cut just below a leaf joint and remove the lower leaves. Cuttings of woody plants often root better if first dipped in hormone rooting powder. Leaf cuttings usually develop roots where the leaf joins the stalk (e.g., saintpaulia), sometimes from the centre vein (e.g., streptocarpus) and in exceptional cases from leaves cut into sections $\frac{2}{5} \times \frac{2}{5}$ in. (1×1 cm) (*Begonia rex* hybrids). Keep the compost fairly dry to prevent the leaves from rotting.

I would advise a rooting mixture of equal parts of sharp sand and peat, which should be kept moderately moist. If you do not own an indoor propagator, a plastic bag tied over the pot (above right) will serve to counteract evaporation.

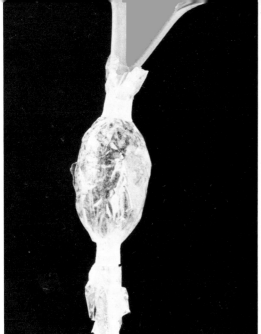

AIR LAYERING

Sometimes a ficus, a monstera or a dracaena loses its lower leaves, and you are left with a bare stem topped by a shock of leaves. If you don't like the plant in this shape, you would do best to resort to air layering, thus taking advantage of the fact that roots can be developed from the stem. The method is as follows: at the desired level, cut off a leafstem close to the main stem. Make a slanting cut halfway into the main stem and open the cut by inserting a matchstick. Moist sphagnum moss (available from a florist) is wound round the cut, and held in place by a sheet of polythene, secured with sellotape. The photograph above left shows the result: a kind of Christmas cracker. The best time of year to do this is June to July, when the plant can be kept fairly warm.

If the moss is kept moist and the soil ball dry, roots will develop in the moss from the cut portion (this takes up to eight weeks). The stem is then severed just below the original cut, the polythene is removed and the upper part of the plant can be carefully potted.

RAISING HOUSE PLANTS FROM SEED

The best way to raise house plants from seed is in a small indoor propagator on the window-sill. Sow from March onwards in small pots filled with a seed compost, three seeds to a pot.

When growth appears, keep the pots in a good light, but *not* in strong sunlight. When the seedlings have two leaves they can be potted out individually in potting compost, but they should be kept in the propagator for the time being. Gradually harden the plants off until they have become accustomed to the drier and cooler room conditions. In May, they can be put on the window-sill.

PLANT CONTAINERS IN THE HOUSE

Since window-sills are often too narrow and, moreover, often have radiators under them, house plants are occasionally placed in containers on the floor. There is no reason why correctly chosen species, with the same requirements for compost, light, temperature, air moisture and water (see pages 113–119) should not be combined in one container. Nevertheless I would use as few different species as possible. Leave the plants in their individual pots, as otherwise the roots will become entangled.

There is no reason why the tubs and urns suggested for use outdoors on the patio should not be used in the home. The fibreglass imitation antique models in white are attractive and light to handle, though they do need a drip tray. The wooden troughs, with or without wrought iron supports, are also acceptable.

PLANTS IN THE OFFICE

Plants can play an important role in making offices seem less stark and uninspiring and more attractive and congenial. In the photograph above right, glass battery tanks, filled with layers of soil in different colours, have been used as plant containers. Other containers and tubs are equally effective. Soilless cultivation, in which plants grow in large containers filled with clay granules, is particularly suitable for offices. The clay granules support the roots which obtain their nourishment from a nutrient solution in the bottom of the container. The level of the solution can be checked and maintained by means of a tube: a practical method, preventing errors in watering.

PRUNING

As a rule you should not prune your house plants too much; you have, after all, bought them to introduce greenery or flowers into the environment, and pruning may spoil the effect. Nevertheless it may occasionally be necessary to remove some leaves and shoots, either because the plant grows too large in the course of time, or because the lower part of the stem may become bare. It is therefore useful to know how to prune.

You should use a sharp knife or secateurs, leaving clean cuts. When a shoot has to be shortened, cut it just above a leaf joint, for it is here that a dormant eye will develop and subsequently produce a new shoot. Sometimes the bud is clearly visible, making it easy to predict the direction of growth of the new shoot.

Plants such as the Chinese rose (hibiscus), which require a resting season in a cooler position, may be cut back rigorously just before dormancy. The two photographs give an example of a bold pruning operation carried out on an abutilon, a species which really does need to be pruned from time to time. The plant will not suffer in any way and will freely produce new shoots in spring.

In some plants, for instance *Euphorbia milii* (the well-known Crown of Thorns), a milky liquid is discharged by the cuts. This 'bleeding' can be stopped with charcoal or with cigarette ash if necessary.

You need not discard all shoots you have removed as in many cases they can be used for cuttings.

SELF-WATERING SYSTEMS FOR PLANTS

If you have no one to look after your plants, holidays can cause problems. Even if you entrust them to someone else, damage may be done. Can we, despite this, leave our plants for a few weeks? Yes, but I would not advise you to place them on bricks in a bowl of water.

The 'mushroom' in the photograph, which sucks up water through a tube and passes it on to the compost via a porous clay peg, works perfectly. Check first how much water the reservoir should contain. An absorbent wick is another useful system. One end of the wick is pushed into the base of the pot through the drainage hole, and the other put into a container of water. Try the system out first to ensure you leave enough water. There are also self-watering troughs and pots, which supply water for a month.

THE PLANT IS ILL

Plants are frequently affected by fungi, aphids or other parasites. Not really surprising, for often our house plants have been 'pepped up' for the market and we are unable to give them ideal growing conditions. In recent years the cultivation of so-called 'disposable' plants has greatly increased; no wonder they cause disappointment if kept too long. Reference is made in the following plant descriptions.

If a hardy plant is attacked by parasites, repotting will nearly always do the trick, since it provides a new source of food. I would not advise spraying with pesticide, as this does not remove the cause of the disease. Too much or too little water, ignoring the dormant season, draught, lack of light, incorrect temperature: all these may result in an attack. Check these factors against the correct conditions described in this book.

Abutilon species

Acalypha hispida

ABUTILON

Easily grown house plants, particularly decorative in their variegated forms. It is important to maintain growth by frequent repotting, as often as three times a year. They may grow to over 12 feet (a good 4 metres) in height. In a sunny situation, or in very good light, flowering is profuse and the foliage is beautifully marked. In winter they tolerate a drop in temperature to about 12°C (53°F). The plants can be cut back vigorously as they always put out new shoots. After a couple of years they grow so large that it is best to grow new plants from cuttings, which will root in a temperature of about 25–30°C (77–86°F) under polyethylene film. In summer, when you should not repot, soak in water (page 10) and feed regularly. Fairly tolerant of dry air.

The photograph above left shows *A. megapotamicum* 'Variegatum' (left) and *A. striatum* 'Thompsonii' (right).

ACALYPHA (Cat-tail)

Attractive house plants, but they do require a moist atmosphere. This makes them difficult to keep, especially in winter. Plenty of light is desirable, but direct sunlight should be avoided. In summer normal moisture should be maintained and the plant should be fed from time to time. Keep warm in winter and spray with water frequently. Standard potting compost. Propagation from tip cuttings—only possible in a heated greenhouse.

The photograph shows *A. hispida*, the Red-Hot Cat-tail, or Chenille Plant. The species *A. wilkesiana*, whose tails are brownish and much shorter, has decorative appeal in the marking of its variegated foliage. 'Marginata' has olive-brown leaves edged with pink, and 'Musaica' produces brown leaves flecked with orange-red and cream.

Achimenes hybrid *Adiantum raddianum*

ACHIMENES (Hot Water Plant)

The name is derived from the Greek and means 'intolerant of cold'. Add to this the fact that it loves humidity, and it will be clear that this is really a hothouse plant. Nevertheless it can be kept in the living-room for quite a long time in summer, but it will need a good deal of feeding and watering. After flowering, it should be kept drier until the foliage has died down. During the winter the scaly rootstock should be kept in practically dry peat at a temperature of 12–16°C (53.5–60.5°F). In April the rootstock is put into fresh compost and started into growth, preferably in a warm greenhouse.

The plants require a great deal of light, but cannot tolerate unshaded sunlight. The achimenes hybrids, which are the ones usually cultivated, vary in colour between pale pink and bluish-purple. Enthusiasts will be able to cultivate a variety of botanical species.

ADIANTUM (Maidenhair)

Particularly decorative ferns, which do not adapt easily to the conditions in a dry living room. They do better when grown with other plants in a container, or, best of all, in a plant window (see page 10) or greenhouse. They have great need of warmth and humidity and should be generously watered and fed. Place in fairly good light, but out of the sun. Towards the autumn, when the plants look less attractive, as is often the case, the old foliage should be removed and, after repotting in potting compost, the plant should be given a few months' rest to bring it into growth again.

There are innumerable species and cultivars of Maidenhair, the most common being *A. raddianum* and *A. tenerum*. Propagation from spores (in the greenhouse).

Aechmea rhodocyanea

Aeschynanthus speciosus

AECHMEA

This is one of the best known bromeliads—that is, each rosette flowers once only, and then dies. However, the plant may continue to flower for up to six months, and will look decorative in the room throughout this time. When buying one, make sure the flower is fresh, for the plant may have been in the shop for months. Situation is not important, since the plant will die in any case. Water moderately; you can pour water into the funnels of the rosette itself. Offsets, which develop at the base of the old plant, can be potted and grown in plenty of (screened) light and a high degree of humidity: temperature 18–22°C (64.5–71.5°F). The potting compost must contain leaf mold, sphagnum moss and peat. The young plant may flower after two or three years, but success is difficult in the living-room.

AESCHYNANTHUS

A little known epiphyte plant ('growing on trees'), originating in the tropical rain forests of southern Asia. It is related to the better known columnea. It *can* be kept in the living room, but if so requires good light, screened from direct sunlight, a temperature of 22–30°C (71.5–86°F) and a high degree of humidity. In other words; hothouse conditions. In winter it may be given a rest, but the temperature should not drop below 18°C (64.5°F). Keeping the soil ball fairly dry during this period will encourage bud formation. In spring repot in a very light mixture consisting of leaf mold, sphagnum moss, well-rotted farm manure, etc. Once it is growing well, water freely. The strongest species is *A. marmoratus*, which has dull green, red-marbled foliage and fairly small flowers. *A. speciosus* is more beautiful, but will thrive only in a greenhouse or plant window (page 10).

Agave americana 'Marginata'   *Aglaonema* 'Parrot Jungle'

AGAVE

In sub-tropical countries these beautiful succulents are often found in parks and gardens. In Western Europe these plants must only be placed outdoors in summer. The most commonly cultivated species is *A. americana* which in some countries is called the Century Plant or American Aloë, as it takes a very long time to flower: the flower stem grows to a considerable length. The finest form is the cultivar 'Marginata' with its yellow-edged foliage, and older specimens may grow to several feet (more than a metre) across. From late May onwards, the plant may be placed in a tub in a sheltered position in the garden, in full sunlight. Do not forget to water and feed it. Bring it indoors at the end of September, place it in a cool position (only just frost-free) and water very sparingly. The potting compost must be porous and calciferous. Propagation from seed.

AGLAONEMA

Little known and fairly delicate foliage plants from Malaysia, not very tolerant of dry living room conditions. Attempts to improve their tolerance by crossing have been partly successful. *A. nitidum* f. *curtisii*, for instance, which has oblong leaves with white marking along the veins, is fairly strong; so is *A. pictum* 'Tricolor', which has smaller, oval and blotchy leaves. 'Parrot Jungle', the result of crossing these two (see photograph above), does well, especially when placed among other plants, which provide some humidity.

Aglaonema likes a high temperature in summer and requires a very light compost (a mixture of sphagnum moss, peat and leaf mold) kept moderately moist. In winter it can stand a drop in temperature to 15°C (59°F). Propagation from tip cuttings or stem sections.

Aloë variegata (left) and *Aloë arborescens*

Ampelopsis brevipedunculata 'Elegans'

ALOË

This plant should not be confused with the agave, which it resembles to some extent, even though the growing conditions are practically the same. The plants belong to different families, and whereas the agave is a native of Central America, the aloë has been imported from South Africa. In summer the plants may safely be placed outdoors, provided the situation is in full sunlight. Keep moderately moist; being succulents, they are satisfied with little water. In winter, these plants can tolerate a drop in temperature to 10°C (50°F) or even lower, as long as they are kept in a frost-free position. Naturally they need even less water at this time.

A beautiful, small species for the window-sill is the Partridge-breasted aloë, *A. variegata* (see photograph, left). In the red container is a shrubby *A. arborescens*, a taller growing species which is fairly common.

AMPELOPSIS

A. brevipedunculata 'Elegans', shown in the photograph, is the only form of this indoor vine in cultivation. It is a climbing or trailing plant suitable for cool rooms. It has a dormant season in winter, when it loses a great deal of foliage. In spring, fine new shoots appear. The plant is most suitable for unheated corridors, draft-free halls, etc. As it is winter-hardy it will not be affected by a few degrees of frost. Like all other variegated plants, this vine needs plenty of light. The compost should be somewhat calciferous. From May till August, the plant is easily grown from tip cuttings, which root readily under glass or plastic at a temperature of 20°C (68°F).

Ananas comosus 'Variegatus'   *Anthurium species*

ANANAS

The variegated pineapple plant, *A. comosus* 'Variegatus', is an ornamental form of the edible species. As with all bromeliads, the parent rosette flowers only once and then dies. If you buy a plant in bud, it may take more than a year for the flower to develop into a small pineapple crowned by a further rosette of leaves. These scarlet-edged leaves are clearly visible in the photograph. Only then will the old rosette deteriorate. If you cut the fruit just below the new foliage and leave it to dry out for a short time, this top section can be rooted in sandy compost.

The variegated pineapple is a particularly strong plant. Older specimens do not require much light for development, but young plants need the best possible light. Water moderately.

ANTHURIUM (Flamingo Plant)

The anthurium needs a great deal of warmth and, above all, humidity—too much for normal living rooms. Nevertheless strong specimens can adapt remarkably well and may then last for years. I have had particularly good results with the large 'Painter's Palette': *A. andreanum*. The common flamingo plant, *A. scherzerianum* (at the rear of the photograph), can also be kept for a long time if sprayed frequently, or if you possess an electric humidifier. The most difficult species is *A. crystallinum* (at the front), which is grown for its foliage.

The potting compost is a mixture of 3 parts leaf mold, 2 parts sphagnum moss, 1 part coarse sand, 1 part rotted farm manure and some small pieces of clay pot. In autumn, the plants should have a resting period at a minimum temperature of 15 °C (59 °F). Give plenty of light, but not direct sunlight. Propagation from stem cuttings.

Aphelandra squarrosa   *Araucaria heterophylla* 'Gracilis'

APHELANDRA

The popularity of these plants has clearly declined. Besides sharing the anthurium's desire for humidity, they need a temperature of at least 18 C (64.5 F), but preferably 22 C (71.5 F), which is essential for their development. They are therefore more at home in a greenhouse or a plant window than on the window-sill above a hot radiator. In the flowering period they need plenty of light but should only be watered moderately. After flowering they should be given a resting period at a minimum temperature of 16 C (60.8 F).

Repot in sandy, nutritious leaf mold. Various strains of *A. squarrosa*, all with variegated foliage, are in cultivation. Propagation is from shoot tips and eye cuttings; this will only be successful in a warm greenhouse.

ARAUCARIA

Although *A. heterophylla* (syn. *A. excelsa*) has always been considered a 'difficult' plant, it is now increasing in popularity. The reason for this may be its great decorative appeal, in spite of the fact that it sheds its lower branches. If you must have a denser effect, you could combine three plants of different sizes in a large tub.

During its growing period, the araucaria likes a temperature of 18 C (64.5 F), but in winter it can tolerate a much cooler position, as long as it is frost-free. Plenty of light is required, but not direct sunlight. Frequent spraying is beneficial, but, in the long run, the trunk will become bare.

Make sure that the compost is acid and give lime-free water.

Ardisia crenata

Asparagus setaceus (left) and *Asparagus densiflorus*

ARDISIA

Attractive small shrubs with magnificent red berries which may last for a long time. In the photograph you can see the berries at the bottom of the plant, while at the top is the inflorescence which will result in next year's crop of berries. To assist the plant in its cycle of producing berries, you should provide plenty of light, with direct sunlight in winter only, and a temperature of 20–22°C (68–71.5°F), a few degrees lower in winter. The atmosphere must be humid, especially when the heating is on. In the flowering period the plant should be kept drier to produce more fruit. The slightly thickened leaf-edges of this *A. crenata* (not to be confused with *Skimmia japonica* which it resembles) contain the bacterium *Bacillus foliicola*, which forms a kind of symbiotic union with the plant. Neither can live without the other. Propagation is from seed; the young seedlings are stopped to encourage branching.

ASPARAGUS (Asparagus fern)

No doubt you know the delicate foliage of *A. setaceus* (left), sometimes added to a bouquet of freesias or carnations. This same foliage can be grown as a house plant. It does not require much light. Summer temperature upwards of 18°C (64.5°F); in winter a dormant season at 8–10°C (46.5–50°F) and very little water. The plant used to be called *A. plumosus*.

To the right of the photograph you can see *A. densiflorus*, better known as *A. sprengeri*, a plant with less delicate foliage, long tendrils bearing thorns and at a more advanced age small white flowers, which may in turn be followed by red berries. This is undoubtedly the stronger species and I have seen it grow in the most unlikely places.

Aspidistra elatior

Asplenium nidus

ASPIDISTRA

The aspidistra is rather an old-fashioned plant, and its renewed popularity, no doubt encouraged by a nostalgia for turn-of-the-century design and decor, is not surprising. The long, leathery leaves are extremely strong and long-lived, the plant will grow at any temperature from 6°C (42.5°F) upwards, will tolerate dry air and requires little light. It even produces small flowers, although you will have to search for them: they appear at the very bottom of the plant, half hidden in the soil. There is also a variegated form, 'Variegata', on the market, but this is less strong and definitely requires more light. The only thing an aspidistra dislikes is a sudden act of kindness on your part in the form of an extra dose of fertilizer. The foliage is unable to cope with such an abundance and will split.

ASPLENIUM

The best known species, and in fact practically the only one now available, is *A. nidus*, the Bird's Nest Fern. Its leaves converge in a rosette. In its native habitat it is an epiphyte (that is, it grows on trees). Although the atmosphere of the tropical jungle is considerably more humid, the plant will nevertheless thrive in our dry living rooms, since the leathery foliage prevents too much evaporation. Regular spraying will increase its beauty. The temperature should not fall below 20°C (68°F). The compost must be light and humusy, and the plant should be kept moist and well nourished.

There are also species with more finely divided foliage, forming young plantlets on their stems. The best known are *A. bulbiferum* and *A. daucifolium*.

Astrophytum species

Aucuba japonica 'Crotonifolia'

ASTROPHYTUM

Among the innumerable cactus species which are grown in greenhouses, a few, because of their distinctive shapes, are popular for growing indoors as well, where they will thrive on a sunny window-sill. The photograph shows *A. ornatum* (left) and *A. myriostigma*, known also as Bishop's Cap. Another well-known species is *A. asterias*, the Sea Urchin Cactus.

One problem which arises when growing cacti indoors is the necessity of keeping them cool in winter—a temperature of 6–8°C (42.5–46.5°F)—and practically dry. This is impossible in the living room, but you can put all your cacti together in a box and place this box in the window of a cool room, or perhaps even in a frost-free garage.

AUCUBA

A small shrub with dark green, leathery leaves, particularly suitable for cool rooms, entrance halls, passages, etc. Since it is really a garden plant, a little frost will do no harm. Too high a temperature, on the other hand, is bad for the plant, especially in winter, and causes it to lose much of its foliage. This plant tolerates shade well, especially the all-green standard form of the plant which is, however, rarely available. The fine variegated strains such as *A. japonica* 'Crotonifolia' (illustrated above) or 'Variegata' with its yellow-flecked green foliage, are more common. Make sure that the compost is acid and never allow the soil ball to dry out entirely. Propagation from cuttings, possibly from seed as well.

26

Azalea indica   *Azalea japonica*

AZALEA

I refer to these beautiful house plants by the popular name by which they are generally known, although they should really be called Rhododendron. The botanical name for *Azalea indica* is *Rhododendron indicum* and the official name for the Japanese azalea, or *Azalea japonica*, which is also grown as a house plant, is *Rhododendron obtusum*.

An azalea is practically always bought in flower. It will create few problems, for a good plant will have plenty of buds as well, which are bound to open provided the plant is not placed in too warm a position. It can be kept in a heated room, but preferably near a window where it is always a little cooler. Water generously and from time to time immerse the pot in water. Wherever possible use rainwater, for azalea does not tolerate lime. The cooler its position, the longer it will flower.

After flowering remove dead blooms, reduce the water supply and keep the plant in good light at 6–8°C (42.5–46.5°F). Remove the first thin shoots that appear. Towards the end of May the azalea is best planted, minus its pot, in a shady spot in the garden; the soil should first be improved with garden peat. Always water well and from time to time feed with a liquid fertilizer. In late September the plant should be repotted in a wide, shallow pot; take care not to damage the roots. Do not put it in a warm room immediately, or it will lose all its leaves; acclimatize it very gradually, and spray frequently with water. In the course of time the buds will open.

The above procedure applies to the indoor azalea, illustrated on the left. The Japanese azalea, on the right, is as a rule winter-hardy and after flowering may therefore be given a permanent outdoor position in acid soil and a damp, half-shady place.

Begonia imperialis   *Begonia maculata*

BEGONIAS—SPECIES

For convenience, we include under this heading all those begonias which do not belong to the commercial assortment of hybrids. They are not grown in such enormous quantities and only appear on the market from time to time. There are approximately 1,000 species, of which 20 to 30 are fairly regularly available. Besides the commercial trade, species begonias in particular are perpetuated by friends and acquaintances privately passing on cuttings. Practically all begonias have in common the property of being simple to propagate, especially from shoot tips or axil cuttings. Even the leaves root easily.

Of the tall, majestic, shrubby forms, the best known are *B. maculata* (illustrated right) and the *B. corallina* hybrids. There are many hybrids which, in large enough pots, may grow about as tall as a man. The large clusters of red, pink or white flowers appear among the big, beautifully marked leaves, mainly in the autumn. Although the plants may last for years, it is advisable to take cuttings every year, for once older plants grow too large to find sufficient nourishment, they may suddenly be affected by mildew and die.

The photograph on the left shows a less upright form, of which there are also many. This is *B. imperialis*, known especially for its velvety brown or emerald-green foliage. *B. manicata* 'Aureomaculata', recognized by the scarlet scales surrounding the leaf-stalk, has particularly fine variegated foliage. *Begonia × erythrophylla* is a very strong hybrid, seen frequently now. It has very large, round leathery leaves and clusters of pink flowers. These are merely a few examples. Creeping species are best propagated from a leaf, complete with its stalk, and including a section of the stem, which will readily root in the well-known sand/peat mixture. The most favourable temperature is 18–22°C (64.5–71.5°F).

Begonia rex hybrid   *Begonia* Rieger type

BEGONIA—WELL-KNOWN COMMERCIAL TYPES

The photograph on the left shows a few leaves of a foliage begonia, *B. rex*, of which innumerable hybrids are in cultivation. These plants give most pleasure when bought in spring and placed in a cool, fairly dark room. If fed and given a little water, they will last throughout the summer. The most favourable temperature is 12–16°C (53.5–60.5°F). If you can provide this in winter as well, keep the plants somewhat drier, cut them back a little and, in spring, repot them in acid compost containing humus. They may then last for another season.

On the right you see a plant which is grown in large quantities, a so-called winter-flowering begonia. These plants now flower at other times of the year as well, but originally flowering occurred mainly in winter. For a long time it was the large-flowered Elatior begonia and the small-flowered Lorraine begonia which were cultivated on a large scale; both were prone to mildew, and stronger types are now preferred, such as the Rieger begonia (photograph right), which is a cross between *B. socotrana* and tuberous begonia strains. These plants, too, can be kept for the longest time at a temperature a little below that of the normal living-room.

Foliage begonias are easily increased by leaf cuttings. Cut the leaves into sections $\frac{2}{5} \times \frac{2}{5}$ in. (1 × 1 cm). Place them flat on the compost, and cover with glass or plastic. They will root within a few weeks. Winter-flowering begonias are propagated from young shoots. Both these methods of growing are more successful in a moderately warm greenhouse, where plenty of light and a high degree of humidity can be provided, than in the living room. The most suitable compost consists of 50% pre-packed potting compost and 50% peat, to which a little well-rotted farm manure, if obtainable, has been added. Do not allow the soil ball to become too dry.

Begonia cucullata hookeri

Tuberous begonia hybrid

BEGONIA—SPECIES FOR HOUSE AND BALCONY

There are two other important groups of begonias which can be grown indoors, on the balcony or in the garden. First, there is *B. cucullata hookeri*, formerly named *B. semperflorens* (illustrated left), a small-flowered form of which various strains have been cultivated for a long time. The flowers are white, pink or red; the foliage may be either green or bronze. In the living room these little plants may flower practically throughout the year, particularly if their position is not too warm. Towards winter they should be given extra light. Cuttings of young shoots can be taken at any time of the year and may root even in a glass of water. These plants are grown in window- or balcony boxes from May until September. They are not fond of direct sunlight.

Tuberous begonias (photograph right) are also sold on a large scale for window cultivation. Here the advice to grow the plants in a cool position is of even greater importance, but this need present no problem in summer if they are placed near a north-facing window that is open from time to time. It would be more difficult in winter, but the tuberous begonia has the advantage that it may die down completely, leaving a tuber to be kept dry at 10°C (50°F). In early spring the tubers are started into growth in damp peat in a box, after which they are repotted in good compost and will flower again throughout the summer—in the garden, if you like.

Finally, a word about the correct pot for growing begonias. I prefer a plastic pot, which does not allow evaporation. This makes it easier to maintain a constant degree of moisture in the soil ball, something which all begonias like. However, you should take care that the compost does not become *too* moist, as so often occurs in plastic pots.

Beloperone guttata *Billbergia × windii*

BELOPERONE (Shrimp Plant)

The decorative appeal of this plant lies in the reddish-brown bracts which grow in drooping spikes. These surround small tubular white flowers which are less significant and soon drop. The plant requires normal living room temperature. It can be kept through the winter at 12–16°C (53.5–60.5°F). Plenty of light and occasional sun will have a favourable effect on the plant's colour. Normal moisture should be maintained in the compost, which can be the standard potting compost. Good ventilation benefits the plant. Cuttings are rooted under glass at 20°C (68°F), after which the plantlets are potted up in groups of 3 to 5. After a few weeks, pinch out the top to encourage bushy growth. Keep the plant on the dry side through the winter to encourage it to rest, and cut back to a third of its size in February.

BILLBERGIA

Billbergia nutans is indestructible, whether placed in a dark or light position. Nor is it very particular about temperature. A curious kind of 'flower' appears among the rosette of leaves; it consists of elongated pink bracts from which droop blue-edged yellowish flowers. After flowering, offsets develop at the base of the plant. These can be used for propagation, but usually they are left with the plant which is repotted as necessary into a larger pot. Finally you have a pot of old and young rosettes, several of which may flower simultaneously. This bromeliad is satisfied with normal humidity and does not need much water. The best potting compost is a light mixture including peat and sphagnum moss. *B. × windii*, a cross between the above species and *B. decora*, has broader leaves, is more robust in all respects, and is equally suitable as a house plant.

31

Bougainvillea spectabilis

Browallia domissa

BOUGAINVILLEA

This plant has become less popular mainly because it is difficult to bring it into flower for a second time. If you have ever seen a bougainvillea in a sub-tropical area, you will know that it requires a great deal of sun and relatively little water to flower profusely. These are the conditions we must provide in July and August when the plant is grown behind glass or in the garden. In winter it has a period of dormancy, during which the temperature may drop to 6°C (42.5°F), and for short periods even below zero. During this time watering should practically cease, although the foliage must be kept moist. Pruning in spring is a mistake, as the plant flowers on the second year's wood. Repot in calciferous leaf mold or potting compost.

BROWALLIA

Browallia domissa (syn. *B. elata*, *B. americana*) is usually raised from seed and sold in flower in May. If you want to sow it yourself, this should be done in February in a heated indoor propagator; the seedlings are pricked out three or four to a pot and stopped once or twice. They are then grown in 5-inch pots (12 cm) and hardened off from late April onwards. Flowering will soon start and may continue throughout the summer. New plants can also be grown from cuttings, which will root readily under glass if given bottom heat.

Flowering will continue longer if the temperature is not too high (12–18°C or 53.5–64.5°F), if there is plenty of light but not strong sunlight and if the plants do not lack nourishment. Ordinary potting compost is suitable.

Brunfelsia pauciflora calycina   *Caladium bicolor* hybrids

BRUNFELSIA

A shrubby plant, flowering chiefly in March and April, after a dormant season from November to January (little water; a temperature of 12–14°C or 53.5–57°F). After this resting period, increase the temperature to 20°C (68°F) and water more liberally. The flowering season is followed by another resting period, during which the water supply should again be decreased. In this period, the plant may be cut back a little and repotted in rich soil mixed with a little peat. Fresh air is essential: the plants may be placed outdoors in summer, but not in full sun. The form most frequently cultivated is *B. pauciflora calycina* illustrated above.

Propagation from half-ripe tip cuttings, which root under glass in a soil temperature of 30°C (86°F), requires patience. From January onwards seeds can be sown in warm conditions.

CALADIUM

As this plant requires a high temperature and, even more particularly, a moist atmosphere, it can only be kept in the living room in summer, when it should be placed in good light, screened from the sun. Water normally and try to maintain maximum humidity. In late September begin to water more sparingly and cease altogether in October. The foliage should then die. Caladium is a tuberous plant and the tubers can survive the winter out of the pot, for instance in dry sand, at a temperature of 18°C (64.5°F). In March they are replanted in rich, humusy potting compost and started into growth in an increased temperature, preferably in a greenhouse. Spray with water frequently.

The forms most often cultivated are *C. humboldtii*, with silver-flecked leaves, and *C. bicolor* hybrids (see photograph), in a variety of shades.

Calathea mackoyana   *Calceolaria* hybrid

CALATHEA

In recent years these foliage plants have proved to be hardier than was at one time believed. Placed among other plants in containers or on an extended window-sill, they will keep fairly well. You will notice from the foliage that they are related to the prayer plant (*Maranta*), but calathea is stronger. In summer it requires a warm position, though in winter it is satisfied with 12–14°C (53.5–57°F). It likes plenty of light, but will grow in a darker position as well. Try to keep the air as humid as possible by frequent spraying. Propagation by division or from tip cuttings.

The plant illustrated is *C. mackoyana*. Other species are *C. lancifolia*, with long, narrow leaves, and *C. ornata*, which has red or white lines between the side veins.

CALCEOLARIA (Slipper Plant)

Once a bestselling plant for Mother's Day, the Slipper Plant has now lost ground to other, more lasting, pot plants. Nevertheless, it is often available at the florist's in April and May. There are large-flowered and small-flowered hybrids for room cultivation in a variety of colours; all have speckled flowers. The small-flowered *C. integrifolia*, always yellow, found among bedding plants, is intended for use on the balcony or in the garden.

Indoors, the Slipper Plant should not have too warm a situation as this rapidly leads to attack by greenfly. Water fairly freely, but without overdoing it, especially if grown in a plastic pot. A high degree of humidity is desirable. After flowering, the Slipper Plant may be cut back and kept through the winter at a temperature of 8–10°C (46.5–50°F), but this is rarely done, as one is unlikely to produce a fine plant in the following season.

Camellia japonica (photo Pat Brindley) *Campanula isophylla*

CAMELLIA

The beautiful *C. japonica* often causes anxiety because the buds drop so easily. It is essential to leave the plant in the same position, for instance in a north-east-facing window, as a change in the direction of the light and an increase in temperature will inevitably cause the buds to fall. The plant likes plenty of light, especially from above. In winter the temperature must never exceed 10°C (50°F)—not easy in the house. When the plant starts into growth at the end of winter, increase the water supply and feed once a fortnight. Repot when growth decreases (June). Use completely lime-free potting compost, such as leaf mold mixed with peat and rotted farm manure. In summer the plant can be put in a sheltered position outdoors. Water (lime-free) frequently.

CAMPANULA (Italian Bell Flower)

I don't think that the Italian Bell Flower (the popular name for *C. isophylla*) is cultivated on quite such a large scale as it used to be. Nevertheless it deserved to be, for it is simple to grow and can give a lot of pleasure, especially to children. It is a perennial and, if cared for properly, will flower profusely for many years.

The plant can be grown in a fairly dark window or in good light, but screened from bright sunlight. Use good-sized pots and rich compost and feed regularly in the growing season. After flowering, the entire plant is cut down to a few inches and placed in good light, in a cool but frost-free position. Water very sparingly. Repot in spring and bring into growth once more; the plant will now be at its best. The blue-flowered species is the most common (see photograph above), but there is a white cultivar, called 'Alba', as well.

Capsicum annuum   *Catharanthus roseus*

CAPSICUM (Red Pepper, Chillies)

You've probably eaten the fruits of this plant at one time or another in the form of chillies or paprika. These are both forms of *C. annuum*; for house cultivation there are multi-coloured strains on the market.

The plants are always grown as annuals. Cultivation starts with sowing under glass in February; extra warmth is essential. The young plants are pricked out, hardened off, stopped once, and finally potted in 4-inch (10 cm) pots. A sunny position is desirable; room temperature is now acceptable. The soil should be kept normally moist. Ordinary potting compost is suitable. Soon the small white flowers appear, followed by the fruit.

In contrast to the solanum which it resembles, there is little point in trying to keep the plant.

CATHARANTHUS

Catharanthus roseus is still sometimes called by its old name *Vinca rosea*, for it is related to a well-known garden plant, the periwinkle or *V. minor*. However, it originates in Java and is therefore not winter-hardy.

The plants, usually sold in flower in summer, are sown in February—in a heated greenhouse, otherwise in a heated indoor propagator. After being pricked out and stopped, they are planted in groups in 4-inch (10 cm) pots. They will flower all through the summer. They can be kept through the winter at about 10°C (50°F), but, needless to say, watering is greatly reduced at this time. In spring they should be repotted in a compost-based mixture and will grow into robust plants. Cuttings may be taken in spring.

36

Cephalocereus senilis   *Cereus species*

CEPHALOCEREUS (Old Man Cactus)
There are a number of these grey-bearded cacti, covered in long snow-white hair as a protection against too fierce sunlight. In Mexico this species grows into columns 50 feet high (15 m) and a foot across (30 cm); they do not flower until they have reached a height of over 16 feet (6 m), so do not expect the plant to produce a cheerful colour in the living room. In contrast to most of the other cacti, this cactus requires a little humidity. Moreover, in winter the temperature should not drop below 15°C (59°F), whereas for other cacti it may fall to below 5°C (41°F). Of course, the plant's life functions continue at this temperature; it should therefore be given some water, and plenty of light is essential. Propagation from seed.

CEREUS (Hedge cactus)
There are several columnar cacti, including *Cephalocereus* mentioned above, as well as *Pachycereus, Selenicereus, Trichocereus*, etc. For the sake of convenience they are often all referred to as *Cereus*. The *C. peruvianus* shown on the right of the photograph is one of the easiest cacti to grow. It will thrive on any sunny window-sill and in the course of a few years will grow so tall that it becomes top-heavy. If given a lot of water and fertilizer it can be forced to the point when it will break the pot. In their native habitat, where they grow more slowly, heights of 50 feet (15 m) are not exceptional. On the left of the photograph you see a cristate form called 'Monstrosus'. This cultivar exists in various forms and also grows rapidly. If you want to see flowers while the plant is still of reasonable size, you should acquire *C. chalybaeus*, which, once it has reached 20 in. (50 cm), produces enormous pink-white blooms. Keep the plants cool in winter.

Ceropegia woodii ssp. *woodii*

Chamaedorea elegans

CEROPEGIA

Ceropegia woodii ssp. *woodii*, formerly *C. linearis* ssp. *woodii*, is an easily cultivated succulent trailing plant, which you could kill only by giving it too much water. It is unlikely that you will give it too little, for the plant can live without water for weeks on end. It will moreover tolerate a warm position in winter, although technically a cool situation is recommended. It does not even require sunlight. All in all, therefore, the easiest of trailing plants; in addition it produces attractive flowers. It is surprising that it is not grown more frequently. Try to get hold of a cutting, plant it in sandy soil and within a few weeks you will have a delightful plant whose tendrils may grow to over 6 feet (2 m). Older plants develop cormlets along the stems; these may also be used for propagation.

CHAMAEDOREA

This small palm is an attractive acquisition for those who do not like large specimens. Its dimensions are such that it can be kept on the window-sill for a very long time. In addition, its branches may be used to good effect in flower arrangements. Another advantage is that the chamaedorea flowers so readily and so early. The photograph shows some of the long umbels covered in small yellow balls; these may be succeeded by berry-like fruits. Like all other palms, this species is grown from seed in a greenhouse. Indoors the plant prefers a temperature of 14–16°C (57–60.5°F), but it can tolerate temperatures outside this range. The soil ball must be kept constantly moist. The compost should contain some lime; if possible mix in some extra clay-based loam. Use pots with extra depth for palms.

Chamaerops humilis

Chlorophytum comosum 'Variegatum'

CHAMAEROPS (Dwarf Fan Palm)

This palm, with its fan-shaped foliage, is very suitable for the living room, for it does not usually grow too large: its maximum height is about 3 feet (1 m). In southern Europe it still occurs wild and may then grow five or six times as tall. In western Europe the plant is increasing in popularity, especially among people with plenty of room and among lovers of tub plants. In summer it can be placed on a sheltered terrace; sunlight is tolerated fairly well, provided there is sufficient water in the tub. From time to time it should be given some fertilizer. In winter, this little palm is best placed in a small, well-ventilated, but frost-free greenhouse, but as long as you can keep the temperature low, it may be kept indoors as well, where it will not require much light. Watering should now be reduced. Repot in spring into deep pots filled with good potting compost.

CHLOROPHYTUM (Spider Plant)

This well-known house plant probably owes its popularity to its rapid growth and to the fact that it keeps so well. There are striped as well as all-green forms. The variegated plants require a little more light, but the all-green ones should not be kept in a dark corner either. The plant is happiest in a hanging pot fixed to the window-frame. It does not require a great deal of heat, but because of its vigour it needs a fair amount of water and feed. Small white flowers will soon appear on long stems; these are followed by young plantlets which can be removed and potted separately. The variegated species, which has white-centred leaves about $1\frac{1}{2}$–2 in. (up to 4 cm) is called *C. comosum* 'Variegatum'. A very similar form is *C. capense* 'Variegatum', which has smaller, white-edged leaves.

Chrysanthemum indicum hybrid

Cissus antarctica

CHRYSANTHEMUM

Few kind remarks can be made about the pot chrysanthemum. It has been turned into an 'all-the-year-round' plant by means of special methods of cultivation such as spraying, blocking out or increasing the light; one might almost say that it can be brought into flower merely by pressing a button. The end product of these 'plant factories' is a disposable plant without any character or scent.

In your living room, this sort of biological time-bomb, chock-full of dangerous chemicals, can only be kept for a short time. You can then transfer it to the garden, but not being hardy, it is unlikely to be successful. The plant, artificially restricted in size, will become much taller when allowed to grow freely.

CISSUS

Together with *C. rhombifolia*, sometimes called *Rhoicissus rhomboidea*, the *C. antarctica* illustrated above is among the most hardy of climbing house plants. The former has smaller, triple leaves. I am constantly struck by the fact that in restaurants, shops, etc., where plants are as a rule not very well looked after, these are often the last survivors in an originally well-filled plant container. They continue the struggle for life in the most impossibly dark positions.

In a fairly cool room the plants will flourish without any problems. In winter the temperature can drop as low as 6–10°C (43–50°F). Calciferous compost is best. Too much light will result in yellowing of the leaves. New plants are best grown from tip cuttings which root easily with bottom heat.

Citrus micrantha microcarpa   *Clerodendrum thomsonae*

CITRUS (Orange Tree)

A small orange tree bearing fruit is not exactly cheap, but will prove to be a most acceptable gift. Do not expect to produce a colorful tree such as you see here from an orange pip. It is possible to grow a small shrub by this method, but it will never bear fruit. The plant in the photograph is a special ornamental form, which no longer has much in common with the ordinary orange.

In winter, this citrus needs a very cool position in the best possible light. Flowering occurs in May, and at this time the plant is best stood outdoors, where it can be pollinated by insects. Indoors the pollen must be distributed with a brush to encourage fruit formation. Keep outdoors as much as possible throughout the summer and give it fertilizer occasionally. Use calciferous potting compost; propagate from cuttings.

CLERODENDRUM

This is actually a hothouse climber, but it can be restricted in form by continually stopping the shoots. It is bought as a flowering plant. To enjoy it for as long as possible you should grow it in a temperature of 18–20°C (64.5–68°F), in good light but out of strong sunlight, and spray it frequently. From the end of November onwards *C. thomsonae* should have a resting period at 12–15°C (53.5–59°F). Water sparingly at this time. It will probably lose most of its foliage. In February, the plant should be cut back a little and repotted in a compost-based mixture. The shoots may grow rather lanky. Propagate from stem cuttings, which root under glass in a temperature of 25°C (77°F). Insert the cuttings three to a $4\frac{1}{2}$-inch (12-cm) pot in a sandy compost.

Cleyera japonica 'Tricolor'

Clivia miniata 'Variegata'

CLEYERA

This relatively unknown plant is often taken for a ficus, but this is quite off the mark, for it belongs to the tea family, and is therefore related to such plants as the camellia. From this fact the more experienced plant lover will rightly deduce that cleyera, too, must be cultivated in a cool situation. The maximum temperature lies between 12° and 20°C (53.5° and 68°F) and, in winter, between 10° and 12°C (50° and 53.5°F), possibly even lower for short periods. Cleyera is best treated as a tub-plant, to be stood outside in summer. If looked after correctly, it may produce isolated, fragrant creamy flowers. The photograph shows *C. japonica* 'Tricolor', a variegated form. When the light is just right, the foliage is touched with red. Standard potting compost. Propagate from tip cuttings, which root under glass at 18–20°C (64.5–68°F).

CLIVIA

You need patience and love to keep a clivia in good health and bring it into flower every year. This is how to achieve success. In summer, during the growing season, keep it out of the sun at about 18°C (64.5°F). Water liberally and feed from time to time. A resting period from October to December is essential with the plant kept at 10°C (50°F) and given less water. Only resume normal watering when the flower stem has grown to 6 inches (15 cm); the temperature may be increased to 15–20°C (59–68°F). After flowering the seed takes 10 months to ripen and new seed-grown plants are ready to flower in three years.

The most suitable potting compost consists of a mixture of leaf mold, clay-based loam, coarse sand and a little dried blood. Repot with care in order not to damage the roots, and do not use too large a pot.

Codiaeum variegatum pictum

Coleus blumei hybrid

CODIAEUM (Croton)

Because this is a popular plant at the florist's you might believe that it is easy to grow. It is not quite so simple. It sells well because it looks so beautiful, but it is difficult to maintain in good condition. A high degree of humidity is essential; frequent spraying with water helps a little, but a humidifier is better. It can be grown successfully in large plant containers, where it is surrounded by other plants which provide some moisture. To maintain its fine colouring, the Croton requires plenty of light. Older plants tolerate sun as well. Throughout the year the temperature should never fall below 18°C (64.5°F). Water normally. Repot in slightly calciferous compost; propagate from cuttings of ripe shoots, which will root in high soil temperature.

COLEUS (Flame Nettle)

This is a sun-loving foliage plant, just right for a south-facing window-sill. Don't forget that it appreciates humidity, however, and spray it regularly. It is advisable to remove the flower spikes, since flowering affects the plant's beauty. It is not easy to keep through the winter; dry air and lack of light causes most of the foliage to drop, or, at any rate, fade badly. Try at least to keep the top shoot alive as this can easily be rooted in spring, even in a glass of water. The plant is also simple to grow from seed, in February, in damp and warm conditions. The best-known forms are *C. blumei* hybrids in a large range of colours from dark red, bright red, salmon pink and yellow, to green. *C. pumilus* is a creeper; it has smaller, dark brown leaves with a pale green edge.

Columnea microphylla   *Cordyline terminalis*

COLUMNEA

It is difficult to say with certainty whether or not columnea is a house plant. In a heated greenhouse it presents practically no problems, but a dry living room atmosphere is very harmful to the sensitive little leaves. Try to spray with water as much as possible, or hang the plant directly above a bowl of water. If you have an electric evaporator or humidifier in your room make sure that the plant can benefit from it.

Columnea is an epiphyte, that is, a plant which in its native habitat grows on trees. Fairly good light is desirable, but bright sunshine is not tolerated. The potting compost must be very light and contain leaf mold and sphagnum moss. The minimum temperature, both in summer and in winter, is 20°C (68°F). Keeping the plant slightly dry in winter encourages flowering. The hardiest forms are the modern hybrids such as *C. × banksii*.

CORDYLINE

The species usually sold at the florist's is *C. terminalis*, often taken for a dracaena. If you look at the stems of older specimens you will agree that they are very similar. In the dry atmosphere of the living room, cordyline unfortunately loses its lower leaves one after another, and, in contrast to the dracaena, this is not attractive. The plant's life span is prolonged by placing it among other plants in a container, or by creating a high degree of humidity. Provide plenty of light, but avoid bright sunlight. Temperature throughout the year, 18–22°C (64.5–71.5°F). Water moderately. Plants which have lost their lower leaves can be air-layered (see page 13). Propagation from shoot tips or stem cuttings.

Crassula species

Crocus

CRASSULA

This is a large genus of succulents of various forms. All species like a warm position in full sunlight in summer, when they should be watered sparingly. Some species flower readily, the more so if during the winter they have cool, dry conditions in a temperature of 10°C (50°F), or a little lower if necessary. In a centrally heated house, a bedroom could provide this temperature; the plants should then be placed near the window.

The photograph shows, front left in the container, the compact *C. perforata* with to the right a *C. × justi-corderoyi* in flower. The tall plant with streaked foliage behind *C. perforata* is called 'Tricolor Jade'. In the centre there is a *C. lycopodioides*, while the tall scarlet flower umbels belong to *C. crenulata*. At the back of the container is the tall *C. arborescens (C. cotyledon)*.

CROCUS

The crocus is not really a house plant, but corms, prepared or not, are popular for indoor flowering, in pebbles, water or compost. To be sure of success, you should buy corms sold specially for the purpose. If they are grown in compost, the pots are buried in a frost-free, not too damp, spot in the garden. Lift them in mid-January to check whether the corms are shooting, and, if they are, bring them into the light in the coolest possible position. Too high a temperature will lead to failure. If the corms are grown in pebbles, the container should be put in the dark in a cool cupboard; do not add too much water to the bowl, and cover the tips of the corms with plastic. Follow the same method for crocuses grown in water containers; keep the water level just below the corms.

45

Crossandra infundibuliformis   *Cryptanthus zonatus*

CROSSANDRA

A semi-shrub originating in Indonesia, which from May to September produces salmon-pink flowers emerging from elongated spikes. It is best cultivated at a minimum temperature of 18°C (64.5°F); plenty of light is desirable, but direct sunlight should be avoided. Adequate humidity, however, is the greatest problem. In a heated room this can only be provided by the use of humidifiers; spraying alone is not sufficient. In winter the temperature may drop to 12°C (53.5°F); the water supply should then be decreased. Timely propagation of new plants from cuttings will give better results than keeping old plants over winter.

Use ordinary potting compost; propagate from cuttings in bottom heat.

CRYPTANTHUS

We could regard these plants as small foliage bromeliads, for their decorative value lies in the fine leaves rather than in the flowers. This fact is clearly illustrated by the photograph. They are terrestrial growing bromeliads, but feel quite at home in a light compost consisting mainly of chopped fern roots and sphagnum moss. The most favourable temperature is 20–25°C (68–77°F), but mature plants will survive for several months at a lower temperature. The same applies to light; in principle the plant should have plenty, but at a lower temperature it is satisfied with less. A mist-fine spray with water at frequent intervals is always beneficial to these attractive rosette plants. About seven species are available, some with variegated foliage, others marked as in the photograph.

46

Ctenanthe lubbersiana   *Cyclamen persicum*

CTENANTHE

This is a little-known member of the maranta family, but in recent years it has increased in popularity, especially for inclusion with other plants. In its native habitat, the plant grows on the edge of open rain forests. There, in a temperature of 18–22°C (64.5–71.5°F), there is adequate humidity, and fierce sunlight is filtered by the foliage of the trees. At night and in winter this plant will tolerate a slightly lower temperature: 18–20°C (64.5–68°F) is sufficient. It is advisable to lighten the compost by the addition of pieces or granules of polystyrene or charcoal. If possible, ctenanthe should be placed, not in an individual pot, but in a plant window or in a hothouse container. If this is not possible, use wide pots, preferably non-porous in order to maintain constant soil moisture. New plants can be grown from tip cuttings rooted in bottom heat.

CYCLAMEN

A flowering cyclamen is best kept in the coolest possible position. In a heated room this would be near a north-facing window, especially one that lets in the cold to some extent. The most favourable temperature is 10–16°C (50–61°F). However, the plant should ideally be given hand-hot water, that is, water of at least 30°C (86°F). Pour a little into the saucer and pour away the excess after 20 minutes. This will suffice until the next day.

After flowering, gradually decrease the water supply until the plant loses all its foliage. Then transfer the corm to a larger pot with new potting compost, and place the pot in a well-ventilated cool position in or outdoors; spray with water frequently. In August, start into growth again, and in October bring into flower in a cool room.

Cyperus alternifolius   *Cyperus diffusus*

CYPERUS (Umbrella Plant)

The Umbrella Plant is one of the easiest house plants available. It will thrive in shade, but will also tolerate plenty of light, provided it is not full sunlight. The temperature may be anywhere between 10° and 25°C (50° and 77°F). In theory, a high degree of humidity is required, but in practice the plant will do very well in a warm room.

One thing must be remembered: the plant must always stand in water. It should therefore be grown in a pot with a deep saucer, or with an ornamental outer pot as in the photograph on the left, with an inch (3 cm) of water in the bottom. This would be fatal for practically all other plants, but is essential for the cyperus.

The photograph on the right shows *C. diffusus*, a low-growing Umbrella Plant which is easily brought into flower. A few spikes are already visible, although this is only a very young plant. Treatment is the same as for the taller growing *C. alternifolius*, so do not forget to stand it in water. If the leaf-tips nevertheless turn brown, they can be carefully cut away, but be sure not to cut into the green part; leave a narrow edge of brown. Naturally, regular spraying with water is excellent.

As the plant grows very rapidly, it should be repotted every year; if there are many new shoots it may even be necessary to do this twice a year.

The plant is propagated by division or by rooting leaf rosettes. The stalks on which these grow should be cut back, leaving only a short section of stem; the rosette is then placed upside down in water. In the course of time, roots will develop and the new plant can be potted.

48

Cyrtomium falcatum

Cytisus × racemosus (photo Harry Smith)

CYRTOMIUM (Holly Fern)

This is an easy fern for indoor cultivation, especially in cooler places such as corridors, halls, bay windows, etc. It will also thrive under the shelf in an unheated greenhouse. The dark green leathery foliage does not require much light, but too high a temperature with the resulting dry atmosphere is harmful. The most suitable growing medium is standard potting compost to which some extra peat has been added. Like all ferns, this species will be most successful if it is frequently manured, enabling it to develop new leaves. The old foliage, which soon becomes unsightly, can then be removed. The species illustrated is *C. falcatum*, a native of Japan, Sri Lanka, China and South Africa. Other, very similar kinds also occur, for instance *C. f. fortunei*.

CYTISUS (Broom)

The once so popular indoor broom is now rarely seen. From time to time the florist may have a supply in stock. The reason for its decrease in popularity is the average high temperature of our living rooms. Broom dislikes such conditions and reacts by dropping its foliage. The most favourable temperature is between 12° and 18°C (53.5° and 64.5°F) and in summer the plant is therefore best put in a sunny spot in the garden. After flowering, but not later than mid-July, it can be pruned a little to keep it compact. In October it should be brought indoors for the winter and placed in good light at 6–10°C (43–50°F); in other words, in the same conditions as cacti. Ordinary potting compost is suitable; it need not be lime-free. Propagation from half-ripe cuttings; these will root under glass at 15–20°C (59–68°F).

Dieffenbachia 'Arvida'   *Dipladenia* hybrid

DIEFFENBACHIA

One plant which has adapted well to our living room conditions is the dieffenbachia. In an old book on house plants which I read recently it was still described as a 'greenhouse plant', but it is now generally realized that it can do very well as a house plant, provided the grower has hardened it off correctly. If brought indoors straight from the hothouse it often does not adapt successfully.

Dieffenbachia must be cultivated in warm conditions, at about 20°C (68°F). In winter the temperature may drop to 15°C (59°F), but you should then beware of giving it too much and, particularly, too cold water. The variegated forms, especially, require fairly good light, but strong sunlight is harmful. In summer, water liberally and feed regularly. Ordinary potting compost may be used. Propagation from tip cuttings, which root at 25°C (77°F) under glass.

DIPLADENIA

These are climbers, with long, twining stems; they originate in tropical regions of America. Growers have never found them to be best-sellers; nevertheless a few hundred thousands find their way to customers every year. Provide a high temperature, especially in spring, and also spray with water constantly, for humidity is essential. In summer, the temperature may be slightly lower. Ensure plenty of light, but avoid bright sunlight. Feed with liquid fertilizer once a fortnight. In winter the plant may be given a rest, but the temperature should never drop below 15°C (59°F). Long shoots should be cut back to 4 inches (10 cm). Repot in spring in leaf mold mixed with peat and sand. The best way to grow new plants is by rooting cuttings at 25°C (77°F).

Dipteracanthus devosianus   *Dizygotheca elegantissima*

DIPTERACANTHUS

An attractive, tropical, semi-shrub from South America, a member of the acanthus family and therefore related to *Fittonia*, among others. Florists and nurserymen usually refer to this plant by its old name *Ruellia*.

The species most often available are the variegated ones, since these are the most attractive. The leaves of *D. devosianus* (illustrated) are marked in white along their veins, and are reddish underneath. In autumn and winter they produce an abundance of small white flowers with blue markings. *D. portellae* resembles the previous species, but has larger, rose-red flowers. Actually these are greenhouse plants, sensitive to dry living room conditions. Temperature, 20–25°C (68–77°F) throughout the year. Avoid bright sunlight. These plants are particularly attractive in plant windows. Propagation from young shoots.

DIZYGOTHECA (Finger Aralia)

This is the narrow-leaved Finger Plant; the broad-leaved kind is *Fatsia*. They are by no means similar in cultivation. Whereas the fatsia must be grown in cool conditions and may even be stood outdoors, dizygotheca definitely requires more warmth, although 20°C (68°F) is perhaps excessive. As a rule the warmer the environment, the drier it is, and this dry atmosphere is something which dizygotheca finds hard to tolerate. It will rapidly lose its lower leaves. The plant is best kept in a slightly cooler position, for instance in a corridor or entrance hall in moderately good light. In the course of many years it can grow very large and decorative. It will grow in standard potting compost. Propagation from seed. Air layering is possible.

Dracaena deremensis 'Warneckii'   *Dracaena godseffiana*

DRACAENA

If you've ever been to the Canary Islands, you doubtless know the Dragon Tree, enormous specimens of which may be found there, some of them a thousand years old or more. Each branch terminates in a bundle of sword-shaped dark green leaves. A red juice can be tapped from the bark; at one time this was used as a colouring agent for paint. Cuttings or seeds are sometimes smuggled into the country. If grown in a temperate, American or European climate this dragon tree (botanical name *D. draco*) will do quite well. It can be treated as a tub-plant; in other words, stood outside in summer and in winter kept at 5–10 °C (41–50 °F), but it may also be grown throughout the year in normal living room temperatures.

Oddly enough *D. draco* is rarely available at the florist's; tradesmen would rather sell you a yucca, which looks somewhat similar, but develops a finer trunk. Cultivation is the same for both these plants.

More commonly available is the species *D. fragrans*, a plant with gracefully curving leaves, growing in a rosette; in the cultivar 'Victoriae' these are striped lengthwise with yellow. In our grandparents' time it was already used as a pedestal plant. *D. deremensis* (photograph left) is slightly stronger; there are a number of cultivars as well. In the course of time the trunk will grow bare, but this will not be unsightly. *D. sanderiana* grows in much smaller rosettes and is always covered all over in leaves. All these species are happiest in a warm room, though on a warm summer day, they like to be put out in a light rain-shower.

D. godseffiana is totally different in appearance; it is more like an ordinary foliage plant. This species flowers readily, but requires far more light than the others.

Dracaena marginata

Duchesnea indica

DRACAENA—continued

I would call *D. marginata*, illustrated above, the strongest house plant in the world. This may not be absolutely correct, for there are a few other indestructible plants, such as *Cissus rhombi-folia* and sansevieria. Nevertheless, this quite delicate looking dracaena belongs to the five hardiest. I have seen it flourishing yards away from a window, and even languishing, but alive, in a perfectly dark corridor, lit only at night by a small lamp. In the long run, all specimens lose their lower leaves, but this problem can be overcome—for instance, by grouping three plants of different sizes in a large container. In sunny weather the plant can be put outdoors; otherwise, keep it in a warm room.

Repot in potting compost, if possible with the addition of some peat and loam. Propagation from stem cuttings.

DUCHESNEA (Fragaria)

You may occasionally find this attractive hanging plant at the florist's. In summer it has yellow flowers, followed by red berries which resemble strawberries, but are quite tasteless. The plant, often referred to as *Fragaria*, presents few problems, provided you grow it in a cool position. In fact, it is happiest in the garden or on a balcony.

Like the common strawberry, the plant will develop runners in the course of summer. These plantlets are particularly hardy; covered with bracken or straw they can spend the winter outdoors. As a rule the parent plant will die. If kept indoors through the winter it should be given a cool situation in good light. Repot in potting compost mixed with extra peat. Propagate from seed in April, or from runners.

Echeveria species

Echinocactus grusonii

ECHEVERIA

Succulent plants from Mexico and the north-western area of South America. Practically all the species form spiralled leaf rosettes. The photograph shows that these rosettes occur in various forms and shades of colour. To obtain most beauty from these plants, give them a very sunny position in summer, possibly outdoors. From October onwards the plants should have a dormant season, during which the temperature may drop to 5–10°C (41–50°F) and the plant should be kept practically dry. In spring, repot in porous, loamy, or at any rate slightly calciferous compost; the temperature should then be increased. During the growing season in summer an occasional rain shower will do no harm. The flowers, most striking in some species, but in others less so, grow on very long stems. Propagation from young rosettes, which will develop in large numbers if the main rosette is removed. The photograph shows (left to right) *E. elegans, E. gibbiflora* and *E. peacockii*.

ECHINOCACTUS

Of this once so extensive genus, *E. grusonii* is practically the sole surviving species, but it is a valuable one. Properly cared for, this cactus with its magnificent yellow spines and numerous ribs, can grow 3 feet (nearly a metre) across. The photograph shows a young specimen. In the United States this cactus does not flower.

To achieve the most vigorous and rapid growth, give the cactus a sunny position in summer, and water adequately. Cover the compost surface with fine gravel (see illustration) to prevent the lower part of the plant rotting as a result of contact with damp soil. Use well-drained pots. Stand in a cool place in winter (5–10°C or 41–50°F) and give little or no water.

Epiphyllum hybrid

Erica × *willmorei*

EPIPHYLLUM

A plant for enthusiasts; usually passed on in the form of cuttings. From the end of May until August, the plant is placed outdoors in a shady position, fed from time to time and watered adequately. In late September bring it indoors, give it not too warm a position and continue to water. From mid-December until well into February, when the flower-buds form, keep it very dry and cool, almost allowing the sections to shrivel up. From now on the plant should not be turned. Gradually give it more water, and spray the leaves. The epiphyllum, or phyllocactus, requires humusy compost, with plenty of leaf mold. To take cuttings, which is easy, cut the joints into 4-inch (10-cm) sections, allow to dry out for a week, and insert upright into sandy soil. Tie the sections to supports to prevent collapse.

ERICA (Heath)

Winter-flowering heath plants are often marketed as house plants. From the end of September onwards you can buy *E. gracilis*, with its profuse clusters of small flowers. This species can be placed outdoors until Christmas, during which time it will flower constantly. It should be fed occasionally. The large-flowered species *E.* × *willmorei* (illustrated) and *E. hyemalis* are more suitable for a cool room or a corridor. The cooler the situation, the longer the plants will last. After flowering, cut back a little and if possible put in the garden in summer. Be careful not to let the soil ball dry out; it is advisable to use plastic pots. As calciferous water is particularly harmful to all heaths and heathers, use unpolluted rainwater if possible. Propagation in January and February or in August, from cuttings grown under glass or plastic at a temperature of 14–18°C (57–64.5°F).

Euonymus japonicus, various strains   *Euphorbia milii*

EUONYMUS

The garden euonymus, known by the popular name of Spindle Tree, produces fruits of unusual shape in autumn. The species used as house plants (and for the garden), which have variegated evergreen foliage, do not develop these beautiful berries. The fact that the many variegated forms of *E. fortunei* and *E. japonicus* (very similar in appearance) are used as garden plants as well, indicates that indoors they should be kept as cool as possible. They are not suitable for a warm room, where they will rapidly wilt. One thing you must ensure is that the soil ball never dries out. In addition a fair amount of light is required. Propagation from cuttings during August to October.

EUPHORBIA MILII (Crown of Thorns)

The Crown of Thorns is a succulent plant with ordinary leaves; it can flower throughout the year. It has notably long, sharp thorns and small red flowers, or rather, red bracts surrounding tiny yellow flowers and growing on stick stems. When you cut the plant, either to take a cutting or to improve the shape, the wounds will exude a poisonous liquid and should be staunched with charcoal or cigarette ash. Loss of foliage indicates the plant's need for a resting period; you should then keep it dry for a time and not increase the water supply until new leaves appear, when you should also give some fertilizer. The plant is happiest in a sunny south-facing window and likes warmth. In winter, it tolerates a drop in temperature to 10–15°C (50–59°F). Propagation from cuttings in summer; the cuttings should be allowed to dry out for a few weeks before being rooted in virtually pure coarse sand.

Euphorbia pulcherrima   *Exacum affine*

EUPHORBIA PULCHERRIMA (Poinsettia)

This striking plant is sometimes referred to as the Christmas rose, but that name properly belongs to *Helleborus*, not now cultivated as a house plant. The poinsettia, or *E. pulcherrima* (illustrated) is now available in a variety of strains with enormous bracts which can last for a whole year. The flowers themselves are very small and inconspicuous. Place the plant in good light and water sparingly. If kept too dry, it may lose its foliage in May. If this should happen, it should be cut back drastically and repotted in normal potting compost. After a month's rest it can be started into growth once again and should then be kept warm and moist. If this succeeds, do not leave the plant in light for more than 10 hours a day from October onwards, otherwise the upper leaves will not take on colour.

EXACUM

This attractive, purple-flowered small plant belongs to the gentian family and is available in flower shops in summer. Its official name is *Exacum affine*. It comes in the category of 'disposable' plants, for there is little point in keeping it after October; in any case, that is practically impossible. Growers sow exacum in February to March, and you could do this yourself in an indoor propagator. Use a mixture of coarse sand, peat and some good loam and maintain a temperature of 20–24°C (68–75°F). Prick out once and a few weeks later plant groups of about 5 little plants in a 4-inch (10-cm) pot, using standard potting compost. The small plants should on no account be kept too warm: 18–20°C (64.5–68°F) in a cold frame is best. The plants will be in full flower about July to August. Place in good light, but out of bright sunlight.

×*Fatshedera lizei*   *Fatsia japonica*

×FATSHEDERA

The cross in front of the name indicates that this is not a natural species but the result of crossing two different genera, in this case *Fatsia japonica* and *Hedera helix,* both described elsewhere in this book.

The characteristics of the fairly popular ×*Fatshedera* are a combination of the two parent plants. It derives its tendency to climb from *Hedera*, but its foliage resembles that of the *Fatsia*. The plant may grow to over 16 feet (5 metres) but usually remains much smaller. Cultivation in a warm room does not suit it any better than *Fatsia* and *Hedera*, but it is happy in only moderately warm corridors, halls, etc. (Young plants do require a little more warmth.) The plant is easily increased from tip or stem cuttings.

FATSIA

This is one of the progenitors of ×*Fatshedera* described above. We generally call it the 'broad-leaved Finger Plant' to distinguish it from the narrow-leaved *Dizygotheca* or Finger Aralia (q.v.). It is sometimes known as *Aralia*, but the correct name is *Fatsia japonica*.

Cultivation is most successful in the coolest possible conditions, 6–12°C (43–53.5 F) in winter. In a heated living room the plant cannot survive. It is so insensitive to cold that in coastal areas it can be grown successfully in the garden throughout the year. In these conditions it will even produce flowers, very similar to those of the shrubby ivy. An all-green fatsia requires little light, but the variegated form 'Variegata' should be placed next to the window to retain the variegations. Ordinary potting compost is suitable. Propagation from seed, but air-layering and growing from cuttings are also possible.

58

Faucaria tigrina

Ferocactus species

FAUCARIA TIGRINA

This is one of those succulent plants which because of their attractive appearance always find buyers. This particular one, moreover, keeps very well, even on the window-sill. In summer, during the growing period, a great deal of water is stored in the thick, white-toothed leaves, which are supposed to resemble a tiger's jaws, so no harm is done if you occasionally forget to water. If, on the other hand, the plant is overwatered, especially in winter, it cannot cope with the excess and will soon rot. In winter *F. tigrina* is best put in good light in a cool position at 5–10°C (41–50°F). When the foliage begins to shrivel, water the plant a little; this will encourage flowering. Propagate from seed or cuttings. Use porous potting soil.

FEROCACTUS

This cactus is known chiefly because of its fine, large spines, in many cases curved like fish hooks. In Mexico, some of the species will grow to 10 feet (3 metres) but in more temperate climates you need not worry about this. The plants grow well from their own roots, but they are sometimes grafted onto the lower stem of a fast-growing species. This has been done with the *F. melocactiformis* (left, in the photograph). Grafting on to a root-stock frequently solves problems of climate. The plant centre front is a *F. latispinus*, while the one rear right is *F. wislizenii*. Like all cacti, they should be grown in warm and sunny conditions in summer, but be kept cool (5–10°C or 41–50°F) and practically dry from November onwards. The most suitable compost is a mixture of leaf mold and loam. Provide good drainage. Propagation from seed, available from specialist cactus growers.

Ficus benjamina

Ficus deltoidea

FICUS

In the course of recent years the ficus has probably become the most important house plant, not only because of its great powers of adaptation, but also, and more especially, because of the numerous species available. Undoubtedly the plant's best characteristic is its tolerance of dry, hot air. This is due to the leathery foliage which prevents evaporation. All the species like a position in a warm room: warmth is essential to their growth. For cool rooms you would do better with other foliage plants: there is plenty of choice (see the surveys, pages 113–119 of this book).

The gracefully drooping plant illustrated left is called *F. benjamina*; in the tropics it is known as the weeping fig. In your living room this exceptionally vigorous shrub could easily reach the ceiling. By that time, the trunk will have developed several aerial roots and the branches will droop like those of a weeping willow. The plant will thrive even at a fair distance from the window. Spray the foliage from time to time with tepid water. Keep indoors even in summer, unless by chance there is a warm rain-shower.

Large specimens are not cheap but worth the price, something which cannot be said of all expensive house plants!

A number of ficus species easily develop fruits, which means that they flower as well (very inconspicuously). One of these is *F. cyathistipula*, with long leaves. The photograph on the right shows a better known species: *F. deltoidea*, still frequently called *F. diversifolia*. The small spoon-shaped leaves are covered in very fine gold spots. This is not the easiest species for room cultivation.

Ficus elastica 'Decora'   *Ficus pumila*

FICUS—continued

The best-known of all ficus species is *F. elastica*, or the rubber plant (photograph left). When cut, white latex is actually exuded. This particular one is the cultivar 'Decora', which has larger, wide-oval leaves. There are white-variegated strains as well, but these are notably more delicate and if placed in too dark a position often revert to green.

The rubber plant does not have to grow straight as an arrow towards the ceiling. If pruned when young, it should branch attractively. Older specimens that have lost their lower leaves can be air-layered successfully (page 13). The equally vigorous and particularly decorative *F. lyrata* (syn. *F. pandurata*) produces even larger, lyre-shaped leaves. Strongly recommended, but more rarely available.

Ficus radicans is a small-leaved, climbing species with pointed leaves. A variegated form occurs as well. The plant illustrated above right is *F. pumila* (syn. *F. repens, F. scandens*), also a climbing or trailing plant, but with miniature leaves, heart-shaped and deeply veined. This plant can cover an entire wall. Also available in a (more tender) variegated form.

All these species require light and porous compost. Standard potting compost is suitable, but you might mix in some extra peat. Good drainage is of the greatest importance. Ornamental pots without a drainage hole are death to the ficus. Water regularly with water at blood heat. Diseases are invariably due to incorrect treatment: too much water, lack of light, too low a temperature. Cut back the plant, repot it and improve the conditions in which it grows. Propagation is fairly simple: cuttings will root readily at any time of the year in a soil temperature of 25°C (77°F).

Fittonia verschaffeltii 'Argyroneura' and 'Pearcei'   *Fuchsia* hybrid

FITTONIA

A beautiful little foliage plant, fairly flat and spreading. The foliage has particularly fine marking. *F. verschaffeltii* has rose-red veins which in 'Pearcei' are bright red. The cultivar 'Argyroneura' has distinctive silvery white veins. These plants do best in a large plant container, placed in a warm position, but out of bright sunlight. They especially like a high degree of humidity. When spraying use tepid water. If you prefer pots, use wide, shallow bowls and a light, humusy compost, the more leafy the better. In summer, during the growing season, feed from time to time. Keep warm in winter as well, but give slightly less water. Take cuttings in spring and root under glass in bottom heat. Young plants should be stopped twice or so to encourage branching.

FUCHSIA

Fuchsias have become very popular again, in recent years, largely due to the exhibitions and lectures organized by clubs. At such events, the numerous forms of fuchsia and the profusion of their flowering inspire fresh enthusiasts. In summer the fuchsia can stand outside, in a very sheltered position. Even better, grow the plant in a lightly shaded greenhouse, maintaining a winter temperature of 10°C (50°F). In room cultivation a south-facing window-sill is best, but the plant must be screened from the sun. In summer it requires a fair amount of water and nourishment. After a resting period in winter, cut the plant back a little and repot in standard potting compost. True enthusiasts use very special mixtures. Young shoots will root easily. Standard and pyramid-shaped plants can be achieved by special pruning methods.

Gardenia jasminoides

Gasteria caespitosa (left) and *Gasteria maculata*

GARDENIA (Cape Jasmine)

This plant is not very common, though the very fragrant, wax-like flowers are still used in buttonholes. Flowering occurs in winter and early spring, after which the plant is drastically pruned and repotted, preferably in a mixture of leaf mold, coarse sand and loam. In summer, it should be kept in full light near a window, sunlight not being harmful. Give fertilizer occasionally, but only up to the beginning of August. Spray the foliage as often as possible. In winter the gardenia should be kept warm, at 18–22°C (64.5–71.5°F); the air must be kept moist. Cut flowers keep reasonably well, especially if put in small glass holders. Propagation from young shoots in January. Soil temperature 30°C (86°F).

GASTERIA

Easy-to-grow succulents, requiring a minimum of water. The thick, fleshy leaves retain sufficient reserves for several months. Once the plant is growing well, it can be watered regularly, but only a little at a time and never between the leaves for this will cause rotting. In good weather the plant can be stood outside in full sunlight from early June until the end of September. In winter the gasteria should be kept cool: 6–12°C (43–53.5°F) is the most favourable temperature range. Stop watering until the foliage shrivels up, then give no more than a few teaspoonfuls at a time. In summer the plant produces small pink or red flowers on very long stems. Propagation from seed, but also from leaf cuttings which should be left to dry out for a few weeks before being rooted in sand.

Grevillea robusta   *Guzmania minor* 'Red'

GREVILLEA ROBUSTA (Silk Bark Oak)

As a house plant, *G. robusta* ought to be extremely vigorous, with strength and resistance to match, for in its native habitat it grows to be very tall indeed. Given a roomy, deep pot, the plant could grow twelve feet high within a few years, but specimens of this size are rare. The most likely reason is that the grevillea thrives in a cool environment. In summer it likes to be outdoors, away from bright sunlight, and in winter a temperature range of 4–6°C (39–43°F) is most favourable. Few houses provide such conditions, but it could be a magnificent plant, for instance, for a virtually unheated stairwell in an apartment house, where it could grow without interference to a height of several floors. Repot in calciferous compost. Place a piece of broken earthenware pot, a crock, over the drainage hole, so that the taproot cannot grow through.

GUZMANIA

A bromeliad genus with a geographical distribution over very large areas of South America. It usually grows as an epiphyte (on trees), but is also terrestrial-growing. There are a number of particularly fine species.

As with all bromeliads, the rosettes die after flowering. By the time you buy the plant it has long ceased growing. This has the advantage that any situation is suitable, but the cooler the position, the longer you will keep the plant. It may last for six months, but by that time it will be very unattractive and should be discarded. The offsets at the base of the plant can be cultivated separately, in a greenhouse or propagator. A high temperature and humidity, and a particularly light compost, give the best growing conditions.

64

Gymnocalycium mihanovichii

Gynura sarmentosa

GYMNOCALYCIUM

These red and yellow globular cacti are now sold on an enormous scale. All too frequently, the eager purchaser thinks he is buying a flowering plant which will be easy to grow, but in this he is mistaken. The globe at the top is not a flower, but a leafless cactus which is not viable on its own. For this reason it is grafted onto a fairly tall segment of a cactus species with powers of assimilation. In summer, the small plants must stand in good light in a warm position, but bright sunlight is not suitable. In winter, they have a dormant season at 10°C (50°F), practically without water. If looked after correctly, the plants will readily put out new shoots. Each little globe can be grafted onto a stock, for which *Trichocereus spachianus* is most commonly used. However, many of these cacti die of lack of proper care, and for this reason they are not often bought a second time.

GYNURA

This plant, which is covered in velvety hairs, changes in colour from green to purple, making an attractive addition to a group of plants. Place the container in good light, to make sure fine shades will occur. The best temperature is 18–20°C (64.5–68°F) throughout the year. Water fairly liberally but carefully as water spilt on the leaves will produce ugly blotches.

In the course of time the plant will become less attractive. It will then develop small flowers, orange-yellow and insignificant. The inflorescence, however, is far from insignificant; it has a horrible smell. For these reasons it is best to grow new plants from cuttings, every four months, which is very easy to do. In this way you will always have fresh, healthy specimens. Use ordinary potting compost.

Haworthia fasciata

Hebe andersonii hybrid

HAWORTHIA

Popular rosette-shaped succulents. Not only are the rosettes beautiful in form, but the marking of the leaves also increases the plant's beauty. The species illustrated is *H. fasciata*, but there are many other types. Species with opaque leaves do not tolerate bright sunlight and need a little more water than those whose leaves have a little transparent 'window' at the top. These are real desert plants; they should be given a rest at 5–10°C (41–50°F) in winter in plenty of light, and watered very sparingly. Species with opaque leaves can be kept in the living room. Pots should be well-drained and be filled with cactus compost or, at any rate, with very porous compost. Several species develop young rosettes at the base; these can be rooted separately after being left to dry out.

HEBE (Veronica)

These charming, profusely flowering plants with green or variegated foliage are sometimes available at florists' or on market stalls. They are *H. andersonii* hybrids, popularly called Veronica. These plants, as well as related species, are often grown outdoors, from which we deduce that indoors they like the coolest possible position. This presents particular problems in the season during which they are sold in flower, namely in autumn: 5–10°C (41–50°F) is the most suitable temperature range. Water fairly freely. After flowering cut back and keep cool in winter. In May, repot in prepacked potting compost and preferably bury the pot in a sunny spot in the garden. On no account forget to water and feed often.

Hebe can be propagated from cuttings at almost any time of the year in bottom heat 20–22°C (68–71.5°F).

Hedera helix   *Hibiscus rosa-sinensis* 'Dan'

HEDERA (Ivy)

The common ivy, *H. helix*, which grows on walls and fences, is also sold as a house plant. The photograph shows a specimen. A number of others with different leaf-shapes are somewhat more interesting, for instance 'Sagittifolia', with small arrow-shaped leaves, and 'Pittsburgh', which has small, deeply incised leaves. There are also variegated strains such as 'Eva', 'Herald' and 'Goldheart'. A subspecies of the common ivy, distinguished by its dull red stems and leaf stalks, is called *H. helix* ssp. *canariensis*. The variegated form 'Variegata', better known as 'Gloire de Marengo', is the one most often cultivated.

Ivy must be grown in cool conditions; it even tolerates frost. The plants rapidly deteriorate in warm rooms. The green species, in particular, are tolerant of shade. An exception is the subspecies *canariensis* which requires a much higher temperature.

HIBISCUS (Chinese rose)

A very popular house plant. The double-flowered strains (see illustration) are now most commonly available, in shades of pink, yellow and orange. They are extremely beautiful, but not as hardy as the good old parent species *H. rosa-sinensis* with its single, red, funnel-shaped flower, now rarely cultivated. At one time specimens of up to 3 feet (1 metre) in height were grown indoors.

The Chinese rose can be positioned in full sunlight. It likes plenty of water and fertilizer in summer, and immersing the pot in water occasionally is also beneficial. In winter, a resting period is advisable; minimum temperature 12°C (53.5°F), little water, but do not let the plant dry out. It can be drastically pruned. Use rich potting compost.

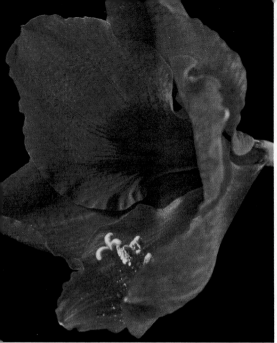

Hippeastrum hybrid   *Howea forsteriana*

HIPPEASTRUM (Amaryllis)

Bringing these bulbs into flower is an increasingly popular pastime. Good bulbs are expensive, particularly if bought by name rather than by colour (named bulbs are obtained from specialist growers and are of high quality).

After potting, as early as November for prepared bulbs, in humusy soil, place in a warm position—for instance, on a not *too* hot radiator. Take great care not to let the compost dry out. When the flower-buds appear, put the pots in good light on the window-sill. After flowering, allow the foliage to develop fully. Feed until August and keep in good light, possibly in a very sheltered position outdoors. From early September onwards, stop watering, and let the foliage die down. Keep bulbs in pots at a minimum temperature of 6°C (43°F). In January top dress the plant and start into growth once more.

HOWEA (Kentia Palm)

H. forsteriana is easy to grow, strong and particularly hardy, developing fairly stiff, broad fans of leaves. *H. belmoreana*, a little more graceful, but also more delicate, requires a slightly higher temperature, especially in winter (15°C or 59°F), whereas *H. forsteriana* is satisfied with 10°C (50°F). In summer the temperature should be about 20°C (68°F). Immerse the pot, rather than water from the top, and every fortnight dissolve a little fertilizer in the water. If you spray frequently and sponge the foliage, the leaf tips may not brown. If browning occurs, however, cut off the tips, leaving a narrow edge of brown along the green. In summer, keep the plant indoors (but put outside in a mild shower of rain). It dislikes full sunlight. Light compost, preferably containing some clay or loam. Use tall pots. Propagation from seed.

Hoya bella (left) and *Hoya carnosa*   *Hyacinthus orientalis* 'Annemarie'

HOYA (Wax Flower)

A good house plant, though it requires warmth (20–25°C or 68–77°F) and humidity, so that it is even better in a greenhouse. In winter, the temperature should not fall below 15°C (59°F). *H. bella* (left), with its little flower umbels and small parted leaves, looks best as a hanging plant. *H. carnosa*, the large wax flower, has oval leaves and large umbels of sweet-scented, pink-centred, waxy flowers. The variegated form, 'Variegata', is very delicate. The main flowering season is May to September. The plants should on no account be turned as this could cause the buds to drop; the stems of dead flowers should not be removed. Plenty of light is essential, but screen from bright sunshine. In the growing and flowering season, water liberally and feed from time to time. Use ordinary potting compost. Propagation from cuttings rooted in bottom heat.

HYACINTHUS (Hyacinth)

Hyacinths can be grown indoors in compost (illustrated), in pebbles or in water. You need special hyacinth glasses for the last method. When growing bulbs in pebbles or in glasses, the water level should always be a little way below the base of the bulb. They are rooted in complete darkness (in a cupboard or cellar) at 9–11°C (48–52°F). Bulbs potted in sand can be buried, frost-free, in the garden. Provide adequate cover, otherwise it is impossible to lift the pots in freezing weather. Transfer hyacinths to light when the flower-bud can be felt. Do not place them in too warm a position. Spray the foliage to prevent drying out. You need specially prepared bulbs for these methods, and after flowering they are useless, unless grown on in good compost. Even so they may miss a season's flowering.

Hydrangea macrophylla   *Hypocyrta glabra*

HYDRANGEA

The indoor *H. macrophylla*, with pink, red or blue flowers, is still available in spring. Placed in a sunny position, the leaves will curl. If this happens, immerse the pot in water and, from then on, keep it in the shade. Water freely and feed from time to time. The temperature should not be too high. After flowering cut back a little, keep somewhat drier and, in May, bury the pot in the garden. In autumn, keep even drier and place indoors in a cool position. In February, in anticipation of flowering, it can be put in a moderately warm room again. The plant is therefore not easy to grow in centrally-heated houses. If left permanently in the garden, it will flower in summer, but in severe winters the frost will destroy the flower-buds of indoor cultivars. To produce blue flowers, add aluminum sulphate to the compost. Propagation from young shoots rooted under glass, May—July.

HYPOCYRTA GLABRA

Introduced fairly recently, this plant is spreading or creeping by nature. The fleshy leaves are small, oval, dark green and shiny. The plant produces small, bulging orange flowers in the leaf axils for much of the year. The best growing conditions are a high degree of humidity and adequate warmth (20–22°C or 68–71.5°F). In a heated room, therefore, the foliage should be sprayed regularly. Sun is undesirable. In December to January, when the days are short, rest the plant at 15°C (59°F), reducing the water supply. At this time the plant can be placed in the sun, if any. The compost must be very light: leaf mold mixed with sphagnum moss and pieces of charcoal. Plastic pots are preferable and must drain very well. Tip cuttings can be rooted in a soil temperature of 20–25°C (68–77°F).

Impatiens wallerana   *Iresine lindenii*

IMPATIENS (Busy Lizzie, Balsam)
Delightful little plants, and easy to grow, they may flower for months on end if left in the same spot in good light. A little sun will do no harm. Flowers of the species *I. wallerana* occur in white, pink, bright red, orange, etc. The leaves are produced in various shades of green and can also be variegated. The stems are succulent, almost transparent. The brighter its position, the more (soft) water it needs. Feed, especially when the plants are in small pots.

Busy Lizzie can be kept through the winter, but it will become very unsightly. New plants are easily obtained from cuttings, which root readily, even in a glass of water. Within a few months, the cuttings will grow into fine flowering plants. Use ordinary potting compost.

IRESINE
This plant used to be included as a non-hardy bedding plant in mosaic designs, the colour of the foliage being very ornamental. The remarkable colouring is caused by an abundance of anthocyanin, a plant pigment which is also dominant in other plants. The more sun, the more striking the colouring. The photograph shows *I. lindenii*, with pointed leaves. *I. herbstii*, which is otherwise very similar, has rather more spoon-shaped, blistery leaves.

Both species are particularly useful to provide contrast in containers with an assortment of other plants, in a sunny position. They are not very long-lived and new plants should therefore be grown regularly from cuttings (very easy). These new plants should be stopped a couple of times.

Ixora hybrid

Jacobinia carnea

IXORA

The beautiful ixora hybrids, whose ancestors came from India and China, often present problems in the living room. To begin with, a change in situation frequently leads to loss of buds, so when you transfer a plant from the grower's greenhouse, you must expect this to happen. It is difficult, moreover, to supply the ixora's need for moist air, especially in a heated room in winter. If you have a greenhouse you of course have no problems, and you can provide the desirable winter temperature of 14–16°C (57–60°F).

In summer, give the plant plenty of light, but *not* bright sunlight. Spray with water often, or even better, buy a humidifier. After flowering allow the plant a month's rest by watering less. Then bring it into flower again. Pot in very light soil, preferably containing leaf mold. Propagation from tip cuttings which root at 25–30°C (77–86°F).

JACOBINIA

This plant may be known to you as *Cyrtanthera*; these changes in nomenclature often lead to confusion. The *J. carnea* illustrated above is the best known species. No doubt you see some relationship with the aphelandra. Both belong to the acanthus family, but there is a distinct difference. Jacobinia needs a temperature no greater than 16°C (61°F). In good light the plant will tolerate more warmth, but good ventilation is certainly desirable. In winter, it tolerates a drop in temperature to 12–14°C (53.5–57°F), and watering should then be decreased. Unfortunately the plant needs high humidity, so that room cultivation is often unsuccessful. From January to April new plants can be grown from young shoots and potted on in standard potting compost. The flowering season is short.

Kalanchoë blossfeldiana

Laurus nobilis

KALANCHOË

The best-known species of the genus, *K. blossfeldiana*, is always forced into early flowering. Keep in the best possible light; the compost must always be moist. A cold frame in the garden is the best place for the plant. When it is growing well, you can give it the 'short-day' treatment. At 8 p.m. cover the plant with a black cardboard cone, then remove this cone next morning at 8 a.m. The most favourable temperature for this treatment, which is started in mid-August, is 15°C (59°F), so it can be done outdoors. When the buds appear, bring the plant into flower normally. In winter, water it less and maintain a minimum temperature of 10°C (50°F). Plants which were formerly called *Bryophyllum* are now included in *Kalanchoë* species. In *K. daigremontiana* and *K. tubiflora* the plantlets appear along the edge of the leaves. They are easily potted on and should be kept warm and in good light.

LAURUS (Laurel, Sweet Bay)

Laurus nobilis is a tree-like shrub from the Mediterranean area. In western Europe it is often cultivated in tubs and shaped like a pyramid or a globular standard tree. The tubs must be filled with a rich, humusy compost and be well-drained. One great problem is that although the shrub is not winter-hardy, it does not tolerate a high temperature in winter either. A well-lit and frost-free garage could possibly serve in place of the old-fashioned conservatory. Temperature 1–6°C (34–43°F); little water. Plants that sprout too rapidly in spring are subject to all kinds of disease. Prune in spring to preserve the shape of the plant. Avoid full sunlight in summer and give some fertilizer occasionally. Propagation from cuttings in autumn or spring; bottom heat 16–20°C (61–68°F).

Mammillaria species   *Maranta leuconeura* strains

MAMMILLARIA

The mammillaria is not the only cactus covered in warts, but, nevertheless, this is a striking characteristic of these attractive, profusely flowering cacti. Some of them flower so readily that they need not be kept cool even in winter (e.g. *M. rhodantha*), and this is quite exceptional for a cactus. In general, however, it is advisable to keep these cacti cool (5–10 C or 41–50 F) and practically dry in winter, with the exception of *M. plumosa*, which flowers in winter and should then be watered. In summer, place in full sunlight and water fairly generously, without over-doing it. The compost must contain loam and sand as well as some leaf mold and peat. Propagation from seed, sprouting species also from cuttings. The photograph shows the species *M. hahniana* (left), *M. lanata* (back right) and *M. spinossissima* (centre front).

MARANTA (Prayer Plant)

The original Prayer Plant is on the right in the photograph. Its correct name is *M. leuconeura* 'Kerchoveana'. Behind it in the photograph is the cultivar 'Fascinator', perhaps even more beautiful, but also more delicate. Having bought these plants with their runners, it is best to transfer them immediately to wide, shallow pots or bowls. Use very humusy soil containing beech leaves and rotted farm manure. The evaporation of water from the moist compost keeps the foliage in good condition. If kept in narrow pots, these fine plants will soon succumb to dry living room air. In summer, keep them out of the sun in a temperature of 18–22 °C (64.5–71.5 °F). In winter the temperature should be practically the same. Propagation by division.

74

Medinilla magnifica

Syagrus weddelliana

MEDINILLA

It is unusual to find these imposing plants, whose splendid flowers touch the soil, in a living room. Nevertheless *M. magnifica* is regularly available at the florist's. Hardly surprising, for these shrubs, characterized by their furrowed stems and the unusual placing of their gutter-shaped leaves, are quite irresistible in their flowering season. Once bought, this plant can only be kept alive if sprayed exceptionally often and given the warmest possible environment, though bright sunlight is taboo. In November to February, the plant should be allowed to rest at 15°C (59°F) and given very little water. In summer, water liberally and occasionally give a dose of liquid fertilizer. Pot in standard potting compost, preferably mixed with leaf mold. Propagation from tip cuttings in a bottom temperature of 30–35°C (86–95°F).

MICROCOELUM (SYAGRUS) (Coconut Palm)

A most attractive small palm from·tropical Brazil, particularly suitable where there is not enough room for larger palms. It must be stressed that its cultivation differs greatly from that of other palms, most of which tolerate low winter temperatures down to 5°C (41°F), while the temperature for the coconut palm should never drop below 18°C (64.5°F), even at night. A further point is that this microcoelum likes a little water in the saucer at all times, something which is fatal for nearly all other plants. Sunlight should be avoided as much as possible. Use standard potting compost and tall, narrow palm-pots which must be well-drained (the beautiful pot in the photograph is not really very suitable).

Monstera deliciosa

Narcissus 'Cragford'

MONSTERA (Swiss Cheese Plant)

The climbing *M. deliciosa*, a foliage plant with unusual holey leaves, has proved to be very tolerant of the dry atmosphere of our living rooms. Many plants grow to quite a reasonable size. Though it will tolerate a lack of light to a point (in fact it dislikes direct sun), it will grow more happily with moderate amounts, and will form more beautiful holes.

A constant room temperature of 20°C (68°F) should be maintained, though the plant will survive at 10°C (50°F). Occasionally the pot should be immersed in water, and timely re-potting into fresh compost is essential. Aerial roots can be trained into the same pot or into other pots, where they will develop into ordinary roots. Older specimens may flower. Propagation from tip cuttings or by air-layering.

NARCISSUS

Bulbs of the 'Paperwhite' or 'Cragford' narcissi, grown in sand or pebbles and put in the living room, can be placed in the light from the start. 'Paperwhite' has pure white flowers; 'Cragford' (illustrated) creamy flowers with orange centres. By setting a bowlful at weekly intervals, mid-August to January you can enjoy the fragrant flowers for weeks on end. The atmosphere should not be too dry. The many other narcissi suitable for forcing should root in the dark at a low temperature (about 10°C or 50°F). This means the bowls should be buried in the garden or put in a cellar. In mid-January, check if the flower-bud has emerged (feel the bulb) and if it has, bring into the light, but in a cool room. After this forced flowering, it is best to discard the bulbs.

Neoregelia carolinae 'Tricolor'

Nephrolepsis exaltata

NEOREGELIA

At one time this plant was called *Aregelia* or *Nidularium*, but the name *Neoregelia* has been used for some time. *Neoregelia carolinae* is the species most often cultivated. Several strains have been developed, green rosettes, all with bright red centres. A very colourful form is the cultivar 'Tricolor' (illustrated), in which the foliage is longitudinally striped with white. The fact that variegated plants are usually less hardy than others need not deter you, since, as with all bromeliads, the rosette is already dying when it starts to flower. Its position is therefore of little importance. Obviously it needs some light and water. The flowers, which are very small, are not very showy, and remain at the bottom of the funnel. Propagation from young offsets, but this is only possible in a greenhouse.

NEPHROLEPIS (Ladder Fern, Sword Fern)

A thriving nephrolepis is always particularly decorative and attractive and, not surprisingly, it is popular for shop-window displays. It is also common in photographic advertisements. But what about the living room, people ask—and with justice, for there the plant is often disappointing because it dislikes dry air. Its life can be prolonged if it is fed weekly to maintain growth. Place in good light (no sun), and be sure to keep definitely moist and not too warm. A clay pot is useless, a plastic one is better. If the plant nevertheless deteriorates, give it a rest of about six weeks by reducing the water. Old leaves are then cut back. After the resting period try to start it into growth again, having first repotted it in rich potting compost. Propagation by division.

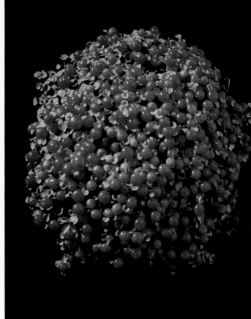

Nerium oleander　　　　　　　　　　　　　　　*Nertera granadensis*

NERIUM (Oleander)

The plant's botanical name is *Nerium oleander*, but it is always called by the second name which sounds more romantic. It is a very common garden shrub in the sub-tropics, and to be successful the plant needs the same conditions over here—that is, a very sheltered and sunny position outdoors in summer, or else a warm window; in winter, a cool environment, 2–10°C (35.5–50°F), though it helps the plant if the pot itself is kept a little warmer. The little water you give it in winter should always be warmed first; and in summer, too, it should never be given cold water. Repot in spring and, if possible, mix some extra loam and a little bonemeal into ordinary potting compost. During the growing season provide a little extra nourishment. Propagation from tip cuttings which will root even in a glass of water, especially if kept in a warm spot.

NERTERA (Bead Plant)

A low-growing plant with tiny, pale green leaves. It is striking because of its innumerable orange berries following the almost invisible small flowers. Place in good light, but out of bright sunlight. Water fairly freely, preferably in the saucer. If you try to bring the plant through the winter, it should not be kept too warm. In August, divide, and repot in sandy leaf mold in well-drained pots. In winter, grow the plants in a cool greenhouse, at about 10°C (50°F), in a high degree of humidity. From March onwards, provide some bottom heat. They soon start into flower, and from then onwards should not be sprayed as this interferes with the setting of the fruit. Pollinate artificially by stroking the flowers with a fine brush. If all goes well, the orange berries will soon appear.

Nidularium fulgens   *Odontoglossum grande*

NIDULARIUM

The name *Nidularium* has already been mentioned on page 77 as a synonym for *Neoregelia*. As the photograph (above left) shows, the true *N. fulgens*, is in fact very similar. In this species the flowers are completely stemless, but you would have to dissect the plant to find this out. A white-striped cultivar of nidularium has existed since 1888, namely *N. innocentii* 'Lineatum'.

Occasionally this strong bromeliad can be propagated successfully. To this end, the offsets are removed when they have grown to half the size of the dying parent plant. The potting compost should contain sphagnum moss and leaf mold. Always keep warm and spray with water as often as possible. From time to time pour a little rainwater into the funnel of the rosettes. The plant may flower after two or three years.

ODONTOGLOSSUM

The odontoglossum illustrated is reputed to be the strongest orchid for room cultivation, because of its great tolerance of dry air. The best position is in an east-facing window, which gets the morning sun. After flowering in the autumn, the plant should be watered very sparingly. Just keep the thick pseudobulbs from shrivelling up. The temperature can drop as low as 8 C (46.5 F). In spring, repot in 2 parts chopped fern roots, 1 part sphagnum moss and 1 part beech-leaf compost. Put a deep layer of broken earthenware pieces at the bottom of the pot. As soon as the plant starts into growth again, water freely, occasionally dissolving a little fertilizer in the water. The first shoots should be sprayed from time to time. Use only rainwater or soft water; hard water is fatal.

Opuntia species

Pachyphytum species

OPUNTIA

The photograph illustrates only a few of the numerous species always available. They should not be handled without gloves, for in addition to the large, sharp spines, they are also covered in bunches of finer spines, the so-called glochids, which are barbed and break off at the slightest touch.

All opuntia species are particularly easy to grow; some will survive outdoors year round in moderate climates (*O. fragilis, O. polyacantha, O. rhodantha*, etc.). All other species can be stood outdoors in summer in a sunny and not too damp position. In winter, they should be kept cool and dry. If grown outdoors, the beds should be well-drained.

It is advisable to use a compost containing a great deal of loam. Most species will not flower until they are very large. The plants are increased by removing sections, which should be left to dry out for some time before they are rooted.

The photograph shows (left to right) *O. pailana, O. bergeriana* and *O. cylindrica.*

PACHYPHYTUM

A small genus of succulents (about 9 species), related to *Echeveria*. The leaves are fleshy and full of sap, and have a beautiful bloom. They should not be touched, as this will damage the bloom. Properly cared for, the plants will flower. In winter, they should have a resting period during which the temperature may drop as low as 2°C (35.5°F). In summer, a sunny position is essential. Do not spray with water; just water sparingly. The leaves drop easily and can be rooted after being left to dry out for some time.

The photograph shows *P. oviferum* (left) and *P. hookeri.*

Pachystachys lutea   Pandanus veitchii

PACHYSTACHYS

This so-called 'modern' house plant has suddenly appeared on the market as if from nowhere. Even the most recent and expensive professional books do not mention the plant. It is, however, referred to in late-nineteenth-century handbooks—it has simply made a comeback. One is immediately struck by its resemblance to the beloperone described on page 31. Pachystachys is not quite so easy to grow and must at any rate be kept indoors throughout the year. The compost must be light and humusy. In summer, water freely and place the plant in good light; be careful with direct sunlight, however. In winter, keep somewhat drier at a temperature of 15 C (59 F). Cut back a little in spring. Propagation from cuttings or stem sections from mid-January until late July.

PANDANUS (Screw Pine)

The leaves of this plant are arranged corkscrew-fashion, hence its popular name. Mature specimens of all species tend to develop stilt-like aerial roots which in their natural habitat can grow to 3 feet (1 metre) in length. In room cultivation you may not notice this phenomenon, but if these aerial roots do become visible, do not cover them with compost. The most favourable temperature is between 18 and 22 C (64.5 and 68 F). Below 15 C (59 F) the plants will soon perish. Plenty of light is desirable, but sunlight should be avoided.

Use roomy pots filled with a light compost (sphagnum, leaf mold, etc.). Water fairly liberally. *P. veitchii* (illustrated) is the most common species, but there is also a species called *P. utilis*, which has red-thorned foliage. Propagation from offsets; *P. utilis* from seed only.

81

Paphiopedilum insigne

Passiflora caerulea

PAPHIOPEDILUM (CYPRIPEDIUM) (Lady's-slipper)

According to many plant books, Lady's-slipper is an excellent orchid for room cultivation, but this is not my experience. Nevertheless these plants are constantly available. The green-leaved species are the least delicate. If you want to give them a try, proceed as follows. Repot in February in a mixture of leaf mold, fern roots and sphagnum moss. Do not give much water to begin with and keep at a temperature of 15 C (59 F). Spray frequently, always using rainwater, and keep out of the sun. In May, when a large root system has been developed, watering should be profuse. Only the foliage should be sprayed.

No actual resting period is required, as these orchids do not have pseudobulbs. In winter, give the plants a fairly cool position, minimum 8 C (46.5 F), and water more sparingly.

PASSIFLORA (Passion Flower)

Various symbols of the life and suffering of Christ are read into the particularly fine flowers of this plant. The most common and strongest species is *P. caerulea*. In summer, the Passion Flower likes plenty of water, and full sunlight does no harm. Carefully tie in the long tendrils. To ensure flowers next year it is essential to keep the plant at 5–10 C (41–50 F) in winter. If you have a sheltered south-facing terrace, plant the passion flower in rich, well-prepared soil against a south wall. For the first few years protect the plant in winter, after which it will be winter-hardy, especially in coastal regions. In time it will grow to several feet in height and width and produce hundreds of flowers followed, in hot summers, by large, orange, oval-shaped fruits. Use potting compost if grown in a container. Propagation from half-ripe shoots, which will root in bottom heat.

Pelargonium grandiflorum hybrid 'Rosalina 390' *Pelargonium zonale* hybrid

PELARGONIUM (Geranium)

It is not surprising that the many forms of the pelargonium (usually called geranium, which is actually the name of a blue-flowering garden perennial) are so popular. They flower readily and because of their succulent tendencies do not mind if one occasionally forgets to water them. Indeed, excess moisture creates problems. *P. grandiflorum* hybrid or French geranium (photograph left) is a true house plant, liking warmth and shelter. It is double-flowered and occurs in a variety of colours.

The much stronger balcony geranium, *P. zonale* hybrid (photograph right), whose leaves have a dark-coloured ring, or cone, likes to live outdoors, in a windowbox or on a balcony. The colour range of these zonal pelargoniums is limitless, but the flowers are fairly small.

Trailing geraniums, *P. peltatum* hybrids, also grow outdoors, on balconies, etc. Cultivation is the same, but their colour range is slightly more restricted. Finally, there are several scented geraniums with incised fragrant foliage to be grown indoors, except during summer.

All geraniums like porous, calciferous soil, such as potting compost with extra sand. In the growing season they should be fed from time to time. As older specimens eventually become unsightly, take cuttings every year in August. Remove the lower leaves from 4-inch (10-cm) tip cuttings, leave the cuttings to dry for a few days, then insert them in very sandy soil. When they have rooted, transfer them to ordinary compost and be sure to keep them dry and cool in winter. The hardy *P. zonale* hybrids tolerate some frost, if they are kept completely dry. Keep the others at a minimum temperature of 5 C (41 F). In March they start into growth once more. If necessary, repot and keep the plants in good light. Sunlight does them good. In May, you should have healthy flowering plants.

83

Pellaea rotundifolia   *Peperomia* species

PELLAEA

These attractive little house plants differ in their cultivation from other fern species in that in their native habitat, they are accustomed to growing on dry rocky slopes. In the United States the *P. rotundifolia* illustrated occurs as far north as Vermont, which indicates that in western Europe, too, it is practically winter-hardy. Dry air presents no problem to the leathery little leaves and, in winter, the pellaea will survive even the unnatural conditions caused by a rise in temperature to 20 C (68 F). Bright sunlight should always be avoided. Hard tap-water is not harmful. This robust, low-growing fern feels particularly at home in a container with other plants. The species *P. viridis* is similar in habit to *P. rotundiflora* but has diamond-shaped leaves.

Propagation by division; the nurseryman grows them from spores.

PEPEROMIA

Although I believe that sales are on the decline, peperomias have been marketed on a large scale for a long time. This is hardly surprising, for, because of their leathery foliage, they are resistant to dry air, the greatest enemy of other house plants. There are a large number of species, not always called by their correct names. The photograph above shows, from left to right: *P. glabella* 'Variegata', *P. caperata* (background) and *P. argyreia*. The photograph on page 85 shows an all-green *P. obtusifolia* on the left, and, to the right, the variegated form 'Greengold', with the species *P. pereskiifolia* behind it. Occasionally the plants do flower, the inflorescence consisting of narrow spikes on long stems, and in some cases, such as in *P. resediflora*, the inflorescence is fragrant.

Peperomia species   *Philodendron scandens*

PEPEROMIA—continued

All peperomia species are natives of South and Central America, where most of them grow in the tropical rain forests, either terrestrially or as epiphytes. Nevertheless, the plants are adapted to living for a time without water, for the upper surface of the leaves is a semi-transparent water reservoir. If you cut across a leaf of a thick-leaved species, such as *P. obtusifolia*, the structure is plain to see. It therefore does no harm if you occasionally forget to water. As a rule, however, the compost should not be too dry. The dark-green species require only a modicum of light, but the variegated forms need a position in good light, out of the sun. Never allow the temperature to fall below 18°C (64.5°F). Use ordinary potting compost mixed with extra peat and plant in wide, shallow pots. Propagation from leaf cuttings and tip cuttings, by division or from seed.

PHILODENDRON

Of the many philodendron species, some of which, like *P. erubescens*, have enormous, heart-shaped leaves, and others (for instance *P. elegans*) deeply incised foliage, we illustrate only the best known: the *P. scandens*. Its staying power is comparable to that of a ficus or monstera. As long as it is given a comfortably warm position, it is practically indestructible. With other, more decorative, species, experiences vary. While one person successfully cultivates the plants with their holey leaves, they may refuse to flourish for someone else. Of course, fairly good light is advisable (but out of the sun). In winter, a minimum temperature of 16°C (61°F). Water moderately and use standard potting compost. Propagation from tip cuttings or eye-cuttings. Air layering is possible.

Phlebodium aureum

Phoenix canariensis

PHLEBODIUM
This fern is still known as *Polypodium*, which is no longer correct. It is a particularly fine house plant, and, in my experience, a very strong one, less capricious than other well-known ferns. The leaf-stems are up to about 2 feet long and the foliage has a beautiful blue bloom. In the cultivar 'Mandaianum' the fronds are crimped. One is guaranteed satisfaction with this fern, not only in an ornamental pot on the window-sill, but also as a hanging plant or combined with other plants. If it ceases to grow for some time, which does happen, keep the plant, with its scaly brown rootstock, dry until it puts out shoots again. When it is growing, always water freely and feed also. The best temperature is between 10° and 21 C (50° and 70°F). Use ordinary potting compost. Propagation by division of rootstock.

PHOENIX (Date Palm)
A truly old-fashioned, rather prickly plant. Nevertheless it is regaining popularity, especially as a useful space-filler in a large hall or stairwell, since it withstands fairly low temperatures. The common date palm, *P. dactylifera*, tolerates as little as 5 C (41°F). A more beautiful species, *P. roebelinii*, has strongly curving fronds but requires a minimum temperature of 14 C (57°F). It also needs a humusy compost, while the others require more loam. *P. canariensis* (illustrated), which is erect in growth, also tolerates quite low temperatures. Use tall, narrow palm-pots. In summer the two hardiest species may be stood outdoors from time to time, but in a very sheltered position out of the sun. Propagation from seed (from a date stone in the case of *P. dactylifera*), but the plants are very slow-growing.

Pilea spruceana 'Silvertree'   *Pittosporum tobira*

PILEA (Artillery Plant)

The pilea owes its popular name to the fact that it discharges its pollen when a flowering plant is moistened. The small plant is very suitable for use in plant arrangements, where it enjoys a higher degree of humidity. Take care not to place it in too dark a position, for its beautiful marking requires plenty of light (*not* bright sunlight). In summer, it may be put outdoors. In winter, the temperature should not drop below 10 C (50 F). Mix potting compost with a little extra peat and coarse sand. *P. cadierei* has dark foliage with light blotches between the veins *P. spruceana* develops blistery bronze-coloured leaves. 'Silvertree' (illustrated) is probably the finest cultivar.

Propagation from tip cuttings, which root even in water.

PITTOSPORUM

A rather less well-known plant with leathery leaves, of a shrubby nature. Some species may remind you of a ficus or a rhododendron, but its cultivation is very different from that of the former, which loves warmth. Pittosporum is a true tub-plant; that is, in summer it likes to stand outdoors (full sun is acceptable, provided you water sufficiently), while in winter, it likes a temperature of 4–10 C (39–50 F). It may also be successful in a sunny window, provided the room is not kept too warm in winter. Pot in ordinary potting compost. With proper care, the plant produces scented white flowers, March to May. The species illustrated is the best known, *P. tobira*, of which there is a variegated cultivar as well. There are several other species on the market. Propagation from tip cuttings, which root under glass in bottom heat in the autumn.

Platycerium bifurcatum *Plectranthus fruticosus*

PLATYCERIUM (Staghorn Fern)

Although sales have declined, the Staghorn Fern will always remain a popular house fern because it is so decorative and rarely disappoints, for the waxy surface of the foliage makes it very resistant to dry air. The foliage bloom should therefore never be rubbed off, for this would lead to the plant's speedy demise. The platycerium prefers a special hanging pot, filled with a mixture of sphagnum moss, fern roots and leaf mold. To water, take the plant down and immerse the pot. Every so often, dissolve a little rotted farm manure in the water. Grow the plant at room temperature, but keep away from strong sunlight. In time, it may develop into a magnificent shape and form black spores on the reverse side of the leaves. Propagation is best achieved by removing young shoots.

PLECTRANTHUS

I cannot say whether this plant is a cure for rheumatism, but some people swear by it and claim they are free from pain as long as they have this plant on the window-sill—preferably in the bedroom. It could be worth trying and it won't cost you anything; if you find a specimen of this rather dull plant (which may grow to quite a size), simply break off a shoot tip and root it in a jar of water. The foliage, and particularly the bluish flowers, show that the plectranthus is related to the better known and more beautiful coleus or flame nettle. It is just as easy to grow. Keep in fairly good light, water quite generously and use ordinary potting compost. Young plants are more attractive, so take cuttings regularly. The species illustrated is called *P. fruticosus*. The species *P. oertendahlii*, which is creeping in habit, has small, circular, crenate leaves, dark green, with lighter veins.

Plumbago auriculata

Polyscias guilfoylei

PLUMBAGO (Leadwort)

An attractive plant, in practice very suitable for tubs, though this is not at first apparent. In summer, it can be placed outdoors in the sun. Give the long shoots some support. Keep compost fairly moist and from time to time give some fertilizer. The plant may flower for months on end. In late September, the whole thing must be transferred to a cool and well-lit winter position. This can create problems. Perhaps you have a frost-free garage? A cactus or fuchsia greenhouse, kept at a temperature of 4–8°C (39–46.5°F), is excellent as well. The soil should now be only just moist. In spring, repot if necessary, using standard potting compost. Propagation from shoot tips in autumn.

POLYSCIAS

This plant is not commonly available at the florist's. Nevertheless, it can be obtained, though large-scale distribution is hampered by the fact that this relative of the dizygotheca, the fatsia and other similar plants, requires a high degree of humidity to flourish. The temperature, too, must be high, 20–22°C (68–71.5°F), but this is less of a problem. Adequate conditions can be created in a peat-filled plant container, in a closed plant window or in a warm greenhouse. The photograph shows *P. guilfoylei*, which has very fine pinnate foliage with a narrow white edge. *P. balfouriana* has three small, round, white-edged leaves to each long leafstalk. If possible use loamy compost, keep the plants out of the sun and water normally. Propagation from tip cuttings at a high bottom temperature.

Polystichum tsus-simense

Primula malacoides

POLYSTICHUM

Several species of this fine, evergreen fern are well-known as garden plants. The species grown as a house plant is *P. tsus-simense*, a native of China and Japan. This is a very strong fern, greatly to be preferred to other, more commonly available, species. Not that it is advisable to grow the plant in a small pot on the window-sill: it is ideal for inclusion with other plants in a large container where the atmosphere is more humid, something which the fern appreciates. Avoid direct sunlight and water very generously. All ferns love being fed, and this one is no exception. The temperature is best kept on the low side—at 10 C (50 F) in winter, if possible. Use ordinary potting compost. Propagation by division when repotting.

PRIMULA

Spring has hardly begun when various species of primula appear in the flower shops. The most graceful is undoubtedly *P. malacoides*, with flowers growing in whorls around the stems (see detail shown in the photograph). In recent years a number of beautiful strains have been developed, but the rose-red forms are still the most common. The foliage and the stems are covered with white powder or farina. The great disadvantage of this primula (and of others as well) is its intolerance of warm air: 10 C (50 F) is really the maximum. You should therefore not buy it for a heated room, but rather for a corridor, a cool entrance hall, etc. Give plenty of water and feed every week. You might try to keep the plant by putting it in a damp, shady spot in the garden in summer, but this species prefers to grow as an annual, so you are unlikely to be successful.

Primula obconica

Primula praenitens

PRIMULA—continued

The photograph (left) shows the well-known *P. obconica*. Its fairly large flowers grow in spherical umbels. The leaves are large and covered in very fine hairs; some of these secrete a substance called primine which can cause an irritation of the skin in allergic people. There are now strains on the market which lack this substance, but unfortunately this fact is not usually indicated on the label. This species tolerates room temperature, but prefers a well-ventilated position. Water and feed during the flowering season. After flowering, remove flower-stems and unsightly leaves. Repot in potting compost and keep a little drier for some time. In summer, the plant can be put in the garden or, better still, in a shady cold frame. Ensure an adequate water supply. In autumn, start giving fertilizer again and transfer the plant to a cool room indoors. If all goes well, the plant will flower again. Yellow edges on the foliage indicate that the soil ball is too dry, or there is an excess of salts. The latter can be improved by repotting.

The photograph (right) shows another popular primula, the species *praenitens*, better known by its old name *P. sinensis*. This species occurs in a variety of beautiful colours and may flower as early as October. It requires maximum light, is fairly tolerant of a warm room temperature and does not need as much water as other species. After flowering, repot in clayey soil, keep cool, possibly in the garden, and feed well.

Other primula species are occasionally available, especially *P. vulgaris*, the primrose sometimes called *P. acaulis*. This species is best put straight into the garden as it is winter-hardy. Damp soil and a shady position. In the living room this plant is permanently thirsty.

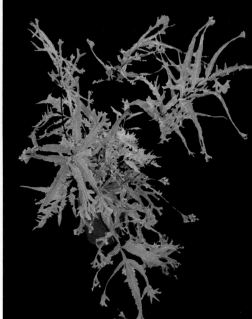

Pseuderanthemum kewense   *Pteris cretica* 'Wimsettii'

PSEUDERANTHEMUM

This relatively uncommon plant is often called by its former name, *Eranthemum*. Most species originate in Polynesia, where of course they grow in tropical conditions. The one usually seen is *P. kewense* (*syn. P. atropurpureum*, illustrated), a species with short-stemmed wine-red leaves, which have rose-pink and olive green blotches. The small white flowers have purple spots. There is also a green-leaved species, *P. reticulatum*, whose leaves are covered in a network of fine golden lines.

Although the pseuderanthemum is actually a hothouse plant, it will survive for a time amongst other plants in plant containers. As a house plant it should be sprayed with water very often, and be grown in a high temperature. The young plants are most beautiful and you should therefore frequently root cuttings (in a bottom temperature of 25°C or 77°F).

PTERIS

About five species of this strong fern are regularly available, but in addition dozens of cultivars of most species have been developed. The choice is therefore enormous and the most extraordinary leaf shapes occur. Most are all-green, but there are also strains with a narrow white edge to the foliage.

The photograph (above right) shows *P. cretica*, a species which can be grown outdoors in Mediterranean countries. Of the innumerable cultivars I have selected the strain 'Wimsettii' as an example. Fairly tolerant of dry living room air.

P. ensiformis, which grows wild in Indonesia and elsewhere, has variously shaped fertile and infertile fronds. The white variegated form, 'Evergemiensis', is well-known.

Pteris tremula

Punica granatum 'Nana'

PTERIS—continued

Another fairly strong species is the shapely *P. quadriaurita*, whose fronds are double-feathered. The side-leaves are sessile—that is, they are stemless. *P. tremula* (photograph left) has delicate foliage resembling that of carrots. It is usually sold fully grown, in fairly large pots.

As all ferns love humidity and fertilizer, it is advisable to provide both. Bright sunlight can never be tolerated. Grow plants in a mixture of potting compost and rotted beech leaves, in well-drained pots. The ferns must be watered liberally. In winter, they can have a very cool situation, provided it is frost-free. They should on no account be kept in a warm living room. Propagation by division; nurserymen grow them from spores.

PUNICA (Pomegranate)

The edible fruit of the common species *P. granatum* is now rarely cultivated for consumption, although both the flesh and the juice have a very pleasant taste. In southern countries the small trees are grown in gardens for their ornamental value; in western Europe they are grown as tub-plants, since they are not winter-hardy.

The low-growing form 'Nana' (illustrated), which never grows beyond 3 or 4 feet (1 to $1\frac{1}{2}$ m) is probably the most suitable for growing in pots. Its cultivation is about the same as *Nerium oleander*; that is, place the plant outdoors in the sun in summer, water adequately, and occasionally give some fertilizer. In winter, it needs a cool position (2–6 °C or 35.5–43 °F) in good light. In May, place it outdoors again. Potting compost containing loam, leaf mold and rotted manure. Do not use over-large tubs. Propagation from leafless shoots in March.

Rebutia turbinata   *Rechsteineria cardinalis*

REBUTIA

This is a genus of miniature cacti with proportionately enormous flowers. The cacti are natives of northern Argentina and Bolivia. They take up hardly any room on the window-sill: a space 40 inches by 8 (100 cm by 20) will accommodate at least 150. The plants will grow in pure sand or in peat. Specially prepared cactus compost is available, and this is, of course, ideal. In summer, they like a sunny position. Water fairly freely and give some cactus fertilizer from time to time. Provide adequate ventilation. In winter, these cacti, too, must be kept cool and dry. It is probably easiest to put them together on a wooden tray and place this tray in the window of a cool room or frost-free garage. Repot early in spring using adequately sized pots. Easily grown from seed.

RECHSTEINERIA

It seems as if these once quite popular plants are gradually disappearing. The plants have tuberous rootstocks, which, like the gloxinia (*Sinningia*), can survive the winter without foliage. In February to March, the rootstocks are planted in light, acid soil, consisting of peat, leaf mold, coarse sand and rotted farm manure, in well-drained pots. The plants will grow rapidly in a high temperature and a moist atmosphere (preferably in a greenhouse). As a rule, the preparatory work is done by the grower and the plants are sold in flower around May. They should then be kept in a warm room out of bright sunlight and must be fed with a very weak solution once a week. After flowering, bring into the sun and reduce the water supply, until all the foliage has died down. During the winter, keep the rootstock in dry peat at about 6 C (43 F).

Rhaphidophora aurea

Rhipsalidopsis gaertneri

RHAPHIDOPHORA

Few florists know this plant's new name so it is still often called *Scindapsus*. No matter; the main thing is the plant's adaptable nature. As long as you maintain a temperature of 18–22 C (64.5–71.5 F) in summer and a minimum of 16 C (61 F) in winter, this climber will grow practically anywhere. The variegated forms require a fair amount of light for in too dark a situation, the plant will revert to green. I have seen rhaphidophora, as well as *Philodendron scandens* and *Cissus rhombifolia*, growing in the most unlikely situations, far from any window. It is advisable to add some peat and sphagnum moss to the standard potting compost. Water moderately and repot in good time. Propagation from cuttings. The species illustrated is *R. aurea*; see also under *Scindapsus*.

RHIPSALIDOPSIS (Easter Cactus)

This cactus is very similar to the Christmas cactus, described under *Zygocactus*. The flowers of the Easter cactus are always bright red and appear at about Easter. After flowering, from May onwards, the plant is best planted, pot and all, in the garden, provided it is in the shade. The soil ball should be kept constantly moist and the plant should be fed every fortnight until early August. In late September bring it indoors once more, place in a fairly cool spot and continue to water. The temperature should not be raised to 5–10 C (41–50 F) until January to February. Keep dry until the buds appear. From now on, do not turn the pot; gradually increase the water supply and place in a warmer position. After flowering the plant may be repotted in ordinary potting compost. Propagation from tip cuttings, which must be left to dry before being rooted.

Rhoeo spathacea 'Vittata'

Rhoicissus capensis

RHOEO

This plant produces small white flowers in shell-shaped bracts, which with a little imagination might be compared to baskets; hence it is sometimes called 'Moses-in-the-Basket'. Despite the resemblance, this plant is not a bromeliad. The rosettes do not die down, but continue to grow and develop offsets. In summer *R. spathacea* requires warmth and the highest possible humidity. The green form tolerates shade, but the variegated 'Vittata' needs much more light, although full sunlight is excessive. Keep the compost moderately moist and feed the plant occasionally. In winter, when a resting period at a lower temperature (up to 10 C or 50 F) is advisable, water the plant less. Use ordinary potting compost. New plants are grown by rooting the offset.

RHOICISSUS

The extremely strong, triple-leaved foliage plant, formerly called *R. rhomboidea*, is now described under *Cissus*. The plant illustrated above is a *R. capensis*, a native of South Africa. Also grown as a climbing shrub, the plant belongs to the vine family. It has beautiful foliage, which, in the young plant, is covered in brown hairs. The plant grows best in a cool greenhouse, as recommended in some plant books, but in my experience it can flourish indoors as well. If it is kept in a cool environment, it does not require too much light. Living room temperature is too much for the plant; moreover, the air is too dry. In a moderately heated hall or corridor, however, the vine grows successfully. Use standard potting compost (which contains more lime) and propagate from cuttings, which root rapidly in some bottom heat.

Rochea coccinea   *Saintpaulia ionantha*

ROCHEA

In recent years I have found it increasingly difficult to obtain a flowering *R. coccinea* (often misnamed *Crassula*). Although it is obviously being cultivated on a smaller scale, it is quite an attractive house plant. It can be kept through the winter, and success is most likely if the plant is not placed in too warm a position. In summer, it can be put in a cold frame, if you have one; otherwise on a cool window-sill, but in good light. After flowering, cut back the stems a little. In summer, water and feed adequately. In the autumn, bring indoors again and keep cool: 4–8 C (39–46.5 F) is best. Propagation from cuttings which root in March; keep the new plants cool. Standard potting compost is suitable.

SAINTPAULIA (African Violet)

It is a remarkable fact that, whereas some people are never successful in growing the African violet, others have no problems in keeping them in flower for 11 months of the year.

If you have had no luck so far, try growing them in flat bowls or dishes, so that all the foliage hangs above the moist soil. Place on a wide window-sill (preventing warm airflow), preferably an east-facing one. A south-facing window must be screened with net curtains or venetian blinds. Now water moderately. Annual repotting is essential (ordinary potting compost is suitable). In summer, feed the plant. This is the best advice I can give.

After flowering, decrease the water supply for a month or so, and keep at 15–18 C (59–64.5 F). Now repot and restart the cycle. Propagation from leaf cuttings: leave a small section of stalk on the leaf, insert halfway into the well-known mixture of 50 % peat, 50 % coarse sand; keep fairly dry.

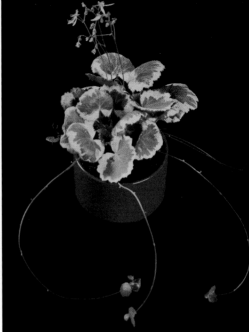

Sansevieria trifasciata and strains

Saxifraga stolonifera 'Tricolor'

SANSEVIERA (Mother-In-Law's Tongue)

Needless to say, this unkind nickname refers to the long-leaved plants at the back of the photograph: left, the common, all-green *S. trifasciata*, and right, its much more attractive cultivar 'Laurentii'. At the front of the photograph are the compact cultivars 'Hahnii' (left) and 'Golden Hahnii' (right). It is impossible to imagine these plants creating any problems. They just dislike temperatures below 14°C (57°F) and excess water left in the saucer. Choose a position where the sun does not penetrate at any time, and the plants will more readily maintain their beauty. Repot in ordinary potting compost only when the plant bursts out of its pot. Propagation of the common all-green species is from 3-inch (8cm) stem sections, which are left to dry before being inserted upright in compost; yellow-edged forms by division only.

SAXIFRAGA (Mother-of-Thousands)

This attractive little plant develops numerous runners, each bearing plantlets, hence its popular name. A Mother-of-Thousands in a hanging pot produces a complete waterfall within a few months. It also provides excellent ground cover in a large container. The common green *S. stolonifera*, the hardiest species, likes low temperatures, about 10–15°C (50–59°F). White-edged 'Tricolor' prefers about 3° more heat, and requires more light. Water sparingly and avoid strong sunlight. Standard potting compost. Propagation method is obvious.

The less common *S. cotyledon* has fleshy rosettes from which emerge tall umbels of small white flowers. This is really a garden plant rather than a house plant, and is better planted outdoors before too long.

Schefflera actinophylla

Scindapsus pictus 'Argyraeus'

SCHEFFLERA

A rapidly growing foliage plant for a moderately warm room. *S. actinophylla* (illustrated) has leathery foliage, with 3–5-lobed, palm-shaped leaves. It will tolerate lower temperatures than *S. digitata*, whose foliage is more divided and much more delicate. As a rule, the temperature should not fall below 12°C (53.5°F) and, in winter, should not rise above 16–18°C (61–64.5°F), which would cause the plant to grow lanky. It will do quite well in a fairly dark position. Water moderately, feed from time to time and provide normal humidity. Use well-drained pots and ordinary potting compost. Propagation from seed sown in a warm greenhouse and the seedlings are then grown on in a moderately warm greenhouse. This plant is strongly recommended for large, fairly cool spaces such as stairwells, entrance halls, corridors, etc.

SCINDAPSUS

The plant generally known by this name is actually a *Rhaphidophora*, described on page 95. The only well-known species now belonging to the genus *Scindapsus* is *S. pictus*, a semi-climbing plant. It is very suitable for plant communities, since it does not need much light, but requires warmth and a fairly high degree of humidity. These conditions are usually present in plant containers. The cultivar 'Argyraeus' has slightly smaller leaves than the species and is more distinctly silver-flecked. Frequent watering is unnecessary, although the soil ball should be constantly moist. Lighten standard potting compost by adding some extra coarse sand and peat, and possibly some sphagnum moss. Use well-drained pots. The plant is easily propagated from cuttings.

Scirpus cernuus   *Sedum* species (see text)

SCIRPUS CERNUUS (Club Rush)

This very beautiful grass-like plant, sometimes called *Isolepis gracilis*, is unfortunately not very common. Initially the stems grow erect, but later they curve gracefully in all directions, giving the impression of a head of green hair. Among the leaf blades are small white balls, which are the inflorescences. If you manage to obtain a specimen of this plant, give it a slightly shady situation. Ensure that the rich, humusy compost remains damp. This is best achieved by leaving the pot in a water-filled dish. The most favourable temperature is between 12° and 20°C (53.5° and 68°F); in winter, 10°C (50°F) will suffice. Moist air is beneficial, but a drier atmosphere is tolerated. When the plant becomes less attractive, it may be divided and the sections repotted separately.

SEDUM

There are numerous sedum species, many of which can be grown in the garden. About ten are known as house plants and are occasionally available at the florist's. Otherwise, they are easily grown from cuttings. In the photograph I have combined three species, namely *S. sieboldii*, a very well-known plant which produces small pink flowers in October (front); *S. griseum*, rear left, a shrubby plant with cylindrical leaves; and *S. rubrotinctum*, rear right, with even thicker, succulent leaves, often beautifully flushed with red. These drop off easily and immediately root themselves.

Sedum sieboldii will grow on the balcony in summer, but the others prefer to remain indoors in a sunny position. They should not be kept too damp. In winter, they must be kept entirely dry at a temperature of 5–10°C (41–50°F).

Selaginella species

Senecio cruentus hybrid

SELAGINELLA (Creeping Moss)

This is a fern, though some species look rather like moss. There are species for the hothouse and others for the cool greenhouse. *S. kraussiana*, for instance, two forms of which are shown in the foreground of the photograph above, grows very well in temperatures varying between 5 and 20°C (41 and 68°F). This species forms clumps, and, in summer, must be watered freely.

Selaginella martensii prefers a slightly warmer environment, in any case not below 12°C (53.5°F). Its cultivar 'Watsoniana', white in parts, is in the background of the photograph above. In general, these plants require more shade than ferns. They need a high degree of humidity, difficult to provide in a heated living room. The pots should be shallow and well-drained. Use rich, light compost. Propagation by division.

SENECIO

You may still know this plant by its more usual, but incorrect, name of Cineraria. The plant illustrated, at any rate, is an *S. cruentus* hybrid, which occurs in numerous colours. It is an annual, which lives longest if kept cool and out of the sun. Keep fairly moist and away from drafts, otherwise it will soon be covered with aphids. Efforts to keep the plant through the winter are hardly worthwhile.

There are succulent forms of the senecio, too, such as the attractive trailing plant *S. mikanioides*, resembling a variegated ivy, but with thicker leaves; and *S. citriformis*, a trailing plant with globular leaves along the stems. Both are easy to grow and may be kept in the living room throughout the year. In winter, place in a slightly cooler position and on no account give too much water. Easily grown from cuttings.

Setcreasea purpurea   *Sinningia* hybrid

SETCREASEA

The best known indoor species is the very hardy *S. purpurea* (illustrated). This plant grows horizontally, which is unusual, and has a fine blue-purple bloom on its leaves. Be careful not to drop water on the foliage, for this quickly causes it to turn unsightly. The small flowers resemble those of the related tradescantia. The plant likes the best possible light and bright sunlight does no harm. In summer, a normal temperature and not too much water, and in winter, a temperature as low as you like. I know someone who has grown the plant outdoors for years, admittedly against a south-facing wall. Take cuttings from house plants from time to time, since this keeps them in better condition.

SINNINGIA (Gloxinia)

You probably know this tuberous plant, now losing favour a little, by the second name. The correct botanical term is 'Sinningia hybrids'. These are divided into a variety of types, different in colour and shape. If you buy a plant in flower, place it in good light, but out of the sun, and give solutions of liquid fertilizer once a week. There is no greedier plant than this one. Neither the foliage nor the flowers should be sprayed. After flowering, place the plant in the sun and gradually reduce the water supply. When the foliage has dried up completely, remove the tuber from the compost and keep in dry peat at about 6°C (43°F). In February, put the tuber in damp peat and keep warm. When the shoots appear, pot the tuber in potting compost with additional peat and keep quite warm. Young leaves may be sprayed with water. Propagation from shoots or leaf cuttings, in bottom heat. New plants may also be grown from seed.

Skimmia japonica

Solanum pseudocapsicum

SKIMMIA

This is really a garden plant, but as it is not completely winter-hardy, it can be a disappointment. If placed in a cool, really well-ventilated room, or better still, in a stairwell, an entrance hall, etc., it will give you much pleasure with a long and healthy life. A large plastic tub is advisable, for the plant dislikes dry soil. At the same time it should not be waterlogged, so ensure good drainage. In summer; it can be planted, pot and all, in a fairly shady spot in the garden. To get berries again, you need to plant a male specimen next to it for pollination. Growers know the male as *S*. 'Fragrans', while the best female form is called *S*. 'Foremanii'. Botanically they are both from *S. japonica*, a dioecious plant. Outdoors, insects carry the pollen from the male to the female plants.

SOLANUM

This plant should not be confused with the orange tree, *Citrus*, or with the Spanish pepper, *Capsicum*. The *S. pseudocapsicum* (Jerusalem Cherry) is a perennial and may survive to quite an advanced age. The plants, purchased in the autumn, are covered in fruit. If then placed in a warm room, the shock is too great and they drop their leaves. This is understandable, for the correct temperature should not be much above 10 C (50 F).

In February, when the fruit has disappeared, they should be cut back slightly and repotted in fresh potting compost. Keep in good light and turn the pot from time to time—this appears to encourage fruit formation.

From the end of May onwards, they are best put outdoors. When in flower, they can be artificially pollinated with a brush. Propagation from seed in February.

Sonerila margaritacea 'Hendersonii'   *Sparmannia africana*

SONERILA

A shrubby plant, suitable mainly for the hothouse or for a well-heated living room with relatively high humidity. In several strains, the foliage is beautifully coloured and marked. The illustration shows *S. margaritacea* 'Hendersonii', whose leaves are marked with silvery-white on deep green; on the reverse side, the leaves are dark purple. The small flowers are rose-red. It is best to transfer the plants to wide and shallow pots immediately after purchase. The damp air arising from the compost is beneficial. Keep in good light, but out of bright sunlight. Maintain normal soil moisture. In spring, grow new plants from tip cuttings, from which the lower leaves have been removed. Soil temperature 30 C (86 F). Put seedlings in a mixture containing beech leaf mold, if possible.

SPARMANNIA (House Lime)

The house lime, of which only a small specimen is illustrated, grows into a tree-like shrub with large, soft, pale green leaves and attractive white flowers during January to April. To prevent its outgrowing the living room, you would be advised to cut it back after flowering to at least half its height. Previous to this, in May, a resting period is good for the plant, during which it may be laid on its side in the garden (not in the sun).

After pruning, repot in prepacked potting compost and bring into growth, either indoors or outside. You should now water generously. In late September the plant must be brought indoors and put in a cool position, preferably not above 10 C (50 F), otherwise frequent spraying with water will be essential. Propagation in April by taking small-leaved cuttings from the flowering shoots.

Spathiphyllum wallisii   *Stenandrium lindenii*

SPATHIPHYLLUM

This hothouse plant can only be kept in the living room for a short time, since it requires a high temperature and great humidity. It resembles the better-known *Anthurium*, but the foliage is much thinner, brighter green and longer-stemmed. The bract is pure white. The plant is very tolerant of shade and will come into flower even if stood under the bench in the greenhouse. Potting compost should be very light; special anthurium compost, containing plenty of sphagnum moss, is best. Put a good layer of broken earthenware crocks in the bottom of the pot. In winter, a drop in temperature to 15°C (59°F) is tolerated, but reduce the water supply then. During the growing season, water fairly liberally, adding a little fertilizer every fortnight. Propagation by division; also from seed, but this takes a long time.

STENANDRIUM

The *S. lindenii* goes well in combination with other plants in a container. It is one of those plants which are useful as a filler for a small space or as ground-cover. In addition, it is very suitable for a terrarium. It is low-growing and bushy. The leaves are dark green and blistered, with yellow marking along the veins. The yellow flower, which resembles that of *Fittonia*, is not the plant's chief ornamental value.

The most favourable temperature for this plant throughout the year is about 20°C (68°F). The air must be as humid as possible, so spray with water frequently or keep the plant under glass. For growing in the living room, plant in shallow bowls filled with potting compost, possibly with some leaf mold added. Propagation from cuttings, which root (needless to say, under glass) at a bottom temperature of 30°C (86°F).

Stephanotis floribunda

Streptocarpus hybrid

STEPHANOTIS (Madagascar Jasmine)

A particularly beautiful twining or climbing plant for the living room or the greenhouse. It can be planted out in the greenhouse; in a pot indoors it is trained along bent wire.

Ideally, the compost should contain a clay loam as well as humus. The plant produces deliciously scented flowers, sometimes used in bridal bouquets, from July to September. Give plenty of light, but keep out of bright sunlight. Water freely and feed only occasionally. In winter, keep drier and cooler—but not below 12°C (53.5°F)—though you should continue to syringe the foliage. Too high a temperature in winter leads to attacks by aphids and mealy bug. Cuttings from the previous year's wood may be taken in spring. Even in a high bottom temperature (25–30°C or 77–86°F), rooting will take a long time (6 weeks).

STREPTOCARPUS (Cape Primrose)

As with the African violet, some people can grow streptocarpus hybrids without difficulty, while others never have success. At any rate the plants should not have too warm a position and never be placed in full sunlight. The foliage should be sprayed with water regularly. In winter when, of course, you give less water, it will do no harm if the temperature drops to 12°C (53.5°F), but do not allow the foliage to die down as in the related *Sinningia* (gloxinia). In spring, repot carefully in potting compost, possibly mixed with a little leaf mold and extra, coarse peat. Place some large broken earthenware crocks in the bottom of the pot. To grow new plants, cut the elongated leaf lengthways along the vein, then insert the cut edge in a mixture of half sand, half peat, preferably warm and under glass. Numerous plantlets develop along the main vein.

106

Syngonium podophyllum 'Albolineatum'   *Tetrastigma voinierianum*

SYNGONIUM

This plant needs almost the same care as the better-known *Philodendron*, to which it is related, and which it greatly resembles. In general, the leaves are much smaller. The hardiest species is probably *S. vellozianum*, whose shiny green leaves have three or five lobes with two curved appendices. The species with grey-green markings, such as *S. podophyllum* 'Albolineatum', are also fairly hardy. In too dark a position, the foliage rapidly reverts to green. They are mainly used in plant communities, where the green species withstand room temperature well. Place the plants in not too dark a spot, spray the foliage regularly and keep the soil ball normally moist. Standard potting compost. Propagation from tip cuttings which only root with bottom heat.

TETRASTIGMA

Of all the members of the Vine family discussed in this book (*Amelopsis, Cissus* and *Rhoicissus*), the cultivation of the *T. voinierianum*—occasionally known as *Cissus*—is the most difficult. This is not saying much, for the others are particularly easy to grow. In my experience, the plants should be left in the same position permanently and the long tendrils should be tied in carefully. Provide plenty of light. If the living room atmosphere is very dry, place the plant in a cooler room, for colder air is always more humid. The temperature may drop as low as 10°C (50°F) and it is, in fact, beneficial if this happens for a period in winter. Then, of course, the plant should be given less water. Tapwater does no harm, as the plant likes lime, and for this reason the compost should contain an alkaline loam. Propagation from cuttings with an eye and one leaf. Do not bury the eye in the compost. Bottom heat 25°C (77°F) and cover with glass or plastic.

Thunbergia alata

Tillandsia cyanea

THUNBERGIA (Black-Eyed Susan)

This climbing plant, usually grown as an annual in western Europe, can be kept through the winter in a greenhouse. It is easily grown from seed in March, in moderate heat. After pricking out once, the young plants are potted and hardened towards early May. For an outdoor position against a wall, be sure that it is south-facing and provide support for the long tendrils (with a few canes, in pot cultivation). The plant also tolerates a good deal of sunlight indoors, provided you give water sufficiently and add some plant food once a week. After flowering, perhaps for several months, the plant may be allowed to grow a little longer outdoors, but in September it must be given a frost-free situation. Growing new plants from seed is much easier. Ordinary potting compost.

TILLANDSIA

This is a relatively less well-known genus of epiphytic bromeliads; in fact, some species grow entirely without roots, for instance the Spanish moss (*T. usneoides*), but this is only successful in a hothouse. The most common species grown for indoor use is *T. cyanea*, which forms a rather stiff rosette with a rose-red inflorescence emerging from the centre. The purplish-blue flowers appear from the bracts one by one. If the plant is still fresh when you buy it, it may produce numerous flowers, provided you give it good light and warmth. The foliage should be sprayed several times a day. After flowering, the pink colour fades and in the course of time the foliage dies down. The plant is now only fit to be thrown away, for I doubt whether you could grow new, flowering specimens from the offsets growing at the base (unless, of course, you have a propagator).

Tolmiea menziesii

Tradescantia albiflora

TOLMIEA (Piggy-back Plant)

A small foliage plant, growing to 8 to 12 inches (20–30 cm) and remarkable because of the plantlets which develop on the mature leaves. A similar phenomenon occurs in some *Kalanchoë* species as well as in *Asplenium*. When the leaf curves towards the soil, roots develop and the plantlet can be potted separately. It is also possible to pot the plantlets before they start into growth. Propagation, therefore, should never be a problem.

Tolmiea does not require a great deal of warmth, 14–18 C (57–64.5 F) is adequate. The compost must be light, such as potting compost mixed with some beech leaf mold. Put broken earthenware crocks in the bottom of the pot to ensure drainage. In winter, the temperature may drop to 5 C (41 F). Never place in too dark a position.

TRADESCANTIA (Wandering Jew)

I cannot tell you why this plant is called Wandering Jew. It is a very common and easily grown house plant. In greenhouses, it grows like a weed and is nearly always found underneath the bench. As a rule, it is cultivated as a hanging plant and if given sufficient light, water and feed, it will develop rapidly. If not looked after properly, the stems grow bare at the base. Don't worry, the shoot tips root easily and you will soon have a new dense-leaved plant plus half a dozen to give away.

Any room temperature between 5 and 25 C (41 and 77 F) is acceptable, but full sunlight should be avoided. Use potting compost. There are several different species, for instance with green and white striped foliage, green foliage, purple underneath, etc. The tradescantia closely resembles the zebrina described on page 112.

Tulipa 'Brilliant Star'

Veltheimia capensis

TULIPA (Tulip)

Several tulip species can be brought into flower indoors. Growers call this 'forcing'. In a good catalogue, species that are easily forced are marked with an asterisk. Tulip bulbs, unlike hyacinth bulbs, are not specially prepared.

Tulips for forcing should always be potted in compost, with the tips of the bulbs a scant couple of inches (4 cm) below the surface. To encourage root formation, it is advisable to bury the pots of bulbs in the garden in early November. Heavily mulch the area where they are buried and, in mid-January, check whether any shoots are visible. This is the time to bring the pots indoors and place them in a cool position on the window-sill. Frequent spraying with water is beneficial. After flowering, forced bulbs can be planted in the garden.

VELTHEIMIA

A bulbous plant originating in South Africa, suitable for cultivation in a moderately warm room. Its flowering season is February to April. The photograph above was taken in April at the Keukenhof in Holland, where I found a number of pots in an attractive, outdoors grouping. To bring into flower, pot the bulbs as early as September, in ordinary potting compost, and keep them cool, about 5–10°C (41–50°F). An unheated greenhouse is the best place. Do not water until the foliage begins to develop. At a later stage, a little fertilizer may be given. Keep in a well-ventilated spot. When in flower, they may be brought into the warm room if you wish. In summer, place them outdoors, so that the foliage will die down. The dormant season is in August and September and, by then, the plants must be very dry. Propagation from offsets, which will not flower for several years.

Vriesea longebracteata

Zantedeschia aethiopica

VRIESEA

These bromeliads are cultivated on a large scale and occur in numerous species, some with red bracts, others with yellow. The flowers themselves, usually yellow in colour, are not very striking. The foliage is often finely marked, especially in *V. hieroglyphica*. The form most often cultivated is the *V. splendens major*, whose correct name is now *V. longebracteata*. The green leaves are cross-banded in brown.

Very occasionally I hear of someone who has succeeded in rooting a young offset and bringing it into flower indoors. This really needs to be done in a warm and humid hothouse. As the old rosette will die in any case, the plant can be placed in poor light. Just water, do not feed. When trying to grow new plants, use a very light compost, containing sphagnum moss and leaf mold or peat.

ZANTEDESCHIA (Calla Lily)

The species of calla lily most suitable for room cultivation is *Z. aethiopica*. The leaves grow fairly large and as this means evaporation on a large scale, the plant should be watered generously, preferably with pure rainwater. The compost must be very rich with a high proportion of humus. After a dormant period in May and June, the plant may be propagated by potting up the small plantlets which have developed in the meantime. In summer, place the plant in the garden and, after the resting period, water and feed it. In September, it should be brought indoors and spend the winter at about 8–10°C (46.5–50°F). If the temperature is higher, you should spray it as often as possible to prevent the flower-buds drying out. Flowering occurs in winter or in spring.

Zebrina pendula 'Quadricolor'

Zygocactus truncatus

ZEBRINA

A hanging plant which greatly resembles several forms of tradescantia. Unlike tradescantia, however, the zebrina does not tolerate temperatures below 12°C (53.5°F), though it can take a much higher temperature. Along its stems the plant has nodes which are clearly visible. At these points, roots develop easily, so growing cuttings in water or moist peat presents no problems. Hang the plant in good light in the sun. This is particularly important for the variegated forms, as they are likely to revert to plain green in a dark situation. *Z. pendula* 'Quadricolor' (illustrated) tends to develop all-green foliage if given too much food or grown in too rich compost, but, correctly cared for, it retains its beautiful markings. The species *Z. purpusii* is more vigorous and has purple foliage.

ZYGOCACTUS (Christmas Cactus)

When not in flower, this cactus is difficult to distinguish from the other leaf cactus, *Rhipsalidopsis*, or the Easter cactus. The flowers are often cerise in colour, but there are also shades of rose-red and red. Flowering occurs about Christmas. After flowering, stop watering and keep cool for a couple of months. Give a little water only when the foliage begins to shrivel. In summer, the Christmas cactus may be placed in the garden and watered freely and fed until early August. Avoid bright sunlight and beware of slugs. Towards the end of September, bring the plant indoors and keep dry and cool. Increase the water supply and the temperature carefully when the flower-buds are visible. Turning the pot or manuring after the beginning of August will cause the buds to drop. Propagation from tip cuttings, which are left to dry out first. Use standard potting compost.

TABLES AND SURVEYS

Which plant where?

Plants are so often placed in the wrong position. A ficus ends up in a cool entrance hall, a schefflera finds itself in a warm room. Within six months they will both be dead, whereas, if their situations had been reversed they would have grown into fine plants in the same period of time.

To save such disappointments, consult the following tables, which should help when selecting a plant for a particular situation or as a gift for a friend.

1. Warm room — full sun

There are very few plants that will survive without problems in full sunlight in a south-facing window, and spend all year in a warm room, but still grow into healthy specimens. The following are some of the species which will tolerate such unnatural conditions, although they, too, will prefer a slightly cooler environment in winter.

Flowering plants

Beloperone
Ceropegia
Euphorbia millii

Hibiscus
Impatiens

Foliage plants

Coleus
Iresine

Sansevieria (will grow unsightly)
Zebrina

---

2. Warm room — screened from daylight

A great many plants like a position which is in good light and yet screened from brightest sunlight. East- or west-facing windows are perfectly suitable for these plants, but a south-facing window must be screened by means of net curtains, venetian blinds or sunshades. A list of plants which may be left in a warm position in this kind of light throughout the year is given below. Please note that, in addition, most of them require humidity, which makes them less suitable for room cultivation. These plants are listed separately, as less easy to grow.

Flowering or fruit-bearing plants, easily cultivated

Ceropegia
Hypocyrta
Pachystachys
Rechsteineria

Saintpaulia
Sinningia
Stenandrium

Flowering or fruit-bearing plants, less easy to grow

Acalypha
Achimenes
Aeschynanthus
Anthurium
Aphelandra
Ardisia

Columnea
Crossandra
Dipladenia
Ixora
Medinilla
Spatiphyllum

Foliage plants, easily cultivated

Asplenium
Calathea
Cyperus
Gynura
Maranta
Pandanus

Pellaea
Peperomia, variegated
Plectranthus
Rhaphidophora
Sansevieria

Foliage plants, difficult to grow

Adiantum
Aglaonema
Caladium
Codiaeum
Cordyline

Ctenanthe
Dipteracanthus
Fittonia
Pseuderanthemum
Sonerila

---

3.   Warm room – shady

Needless to say there are limits to the growing conditions which plants will accept. A reasonable amount of daylight is essential if the plant is not to die. A 'reasonable' amount can be expressed only in *lux*: 1,000 to 1,500 lux is about the minimum. The very strongest plants, most of which are probably familiar to you since they are found everywhere, are listed below. They like warmth throughout the year. With one exception they are all foliage plants, although some may occasionally produce small flowers.

Flowering plants

Billbergia

Foliage plants

Asparagus
Aspidistra
Cissus rhombifolia
Dieffenbachia
Dizygotheca
Dracaena
Ficus, most species
Microcoelum

Monstera
Peperomia, dark green forms
Philodendron
Phlebodium
Platycerium
Rhaphidophora
Sansevieria
Syngonium

---

4. Cool winters – plenty of light

Many plants dislike a winter temperature as high as 20°C (68°F). They need a rest and this is only possible at a somewhat lower temperature. The correct temperature is generally given in the individual plant descriptions. The winter temperature required by the plants listed below varies between 5° and 15°C (41° and 59°F). A hall, a cool bedroom, a corridor, a stairwell— all these are suitable situations.

In summer these plants naturally tolerate normal temperatures, even if, on occasion, these rise to 30°C (86°F). Many of them may also be stood outdoors. All the plants mentioned require plenty of light, but this does not mean that they can all be put in full sunlight. Succulents and cacti are listed separately.

Flowering plants and those bearing decorative fruits

Abutilon
Araucaria
Bougainvillea
Catharanthus
Ceropegia
Citrus
Cyclamen
Cytisus
Erica
Hydrangea
Impatiens
Jacobinia

Kalanchoë
Nerium
Nertera
Passiflora
Pelargonium
Pittosporum
Plumbago
Primula
Punica
Rochea
Solanum
Veltheimia

Foliage plants

Ampelopsis
Chlorophytum
Cleyera
Cyperus
Duchesnea
Euonymus
Grevillea
Hebe
Iresine
Laurus
Nephrolepis

Pilea
Plectranthus
Rhoeo
Saxifraga
Scirpus
Selaginella
Setcreasea
Sparmannia
Tetrastigma
Tolmiea
Tradescantia

Succulents

Agave
Aloë
Crassula
Echeveria
Faucaria

Gasteria
Haworthia
Pachyphytum
Sedum

Cacti

Astrophytum
Cephalocereus
Cereus
Echinocactus
Ferocactus

Gymnocalycium
Mammillaria
Opuntia
Rebutia

---

5. Cool winters — limited light

In cool corridors, halls, porches, etc., there are usually places where little light penetrates. Provided these are not too dark, the plants listed below should survive in such a situation. Minimum temperatures vary between 5° and 12°C (41° and 53.5°F). Many of these plants may be stood outdoors in summer—some, in fact, throughout the year.

Flowering plants

Azalea
Chamaedorea

Zantedeschia

Foliage plants

Aspidistra
Aucuba
Chamaerops
Cissus antarctica
Cyrtomium
× Fatshedera
Fatsia
Hedera

Howea
Phoenix
Polystichum
Pteris
Rhoicissus capensis
Schefflera
Skimmia
Tradescantia

6. Hanging and climbing plants

Plants with a distinct hanging (H) or climbing (C) habit are listed below. The hanging plants are best grown in special hanging pots or baskets. The climbers require a strong cane, or better still, some nails in the wall, to which they are tied at intervals.

These plants have also been mentioned in other sections.

For a warm situation throughout the year

Asparagus (H)
Ceropegia (H)
Columnea (H)
Ficus pumila (H, C)
Ficus radicans (H, C)

Hoya (C)
Monstera (C)
Stephanotis (C)
Syngonium (C)

To be kept cool in winter

Ampelopsis (C)
Campanula (H)
Chlorophytum (H)
Cissus (C)
Duchesnea (C)

Hedera (C)
Passiflora (C)
Saxifraga (H)
Tetrastigma (C)
Thunbergia (C)

7. Plants for tubs

In former times country-house owners cultivated subtropical plants in large wooden tubs. In winter these were transferred to the orangery and kept in good light and certainly frost-free. A number of these plants are still available as house plants. You cannot give them greater pleasure than by treating them as tub plants. Finding a cool but frost-free winter situation may present some problems. It might be possible to adapt a garage (for instance, by putting a window in the roof) so that it is made suitable for the purpose. It need not be warm. In severe frost one might introduce a small electric radiator.

Agave
Aucuba
Bougainvillea
Citrus
Cleyera
Euonymus

Nerium
Pittosporum
Plumbago
Punica
Skimmia
Solanum

8. Disposable plants

The plants listed below can be enjoyed for a short time only. Either they are impossible to keep, or it is not worthwhile attempting to do so. However, they usually look most attractive when you buy them.

If you dislike the idea of throwing away plants, you would do better not to buy these. If you do not mind plants whose ornamental value is only temporary, those listed below come within this category.

Aechmea
Ananas
Browallia
Calceolaria
Capsicum
Chrysanthemum
Crocus
Cryptanthus
Euphorbia pulcherrima
Exacum

Guzmania
Hyacinthus
Narcissus
Neoregelia
Nidularium
Senecio
Thunbergia
Tillandsia
Tulipa
Vriesea

9. Plants for the true enthusiast

If you want to give a plant to someone known to be a real plant lover, you may be sure that a specimen from the category of 'disposable plants', Table 8, will not be satisfactory. So what to buy?

Below you find a number of plants which will give their owners pleasure for years to come. They are not without problems in cultivation, since most of them have a special growth rhythm, but they may thrive in the living-room, and this is after all the main thing.

Begonias, especially
   the botanical forms
Brunfelsia
Camellia
Campanula
Clerodendrum
Clivia
Cyclamen
Epiphyllum
Fuchsia
Gardenia

Hibiscus
Hippeastrum
Hoya
Odontoglossum
Passiflora
Rhipsalidopsis
Saintpaulia
Stephanotis
Streptocarpus
Zygocactus

INDEX OF PLANTS

121

127